Solid Waste Management–
An Indian Perspective

Edited by

M.S. BHATT
ASHEREF ILLIYAN

Synergy Books India
(A Unit of M/s Bookwell Publication)
New Delhi

Solid Waste Management: An Indian Perspective

First Published, 2012

ISBN:978-93-82059-05-9

Published by:
Synergy Books India
(A Unit of M/s Bookwell Publication)
24/4800 Ansari Road, Darya Ganj
New Delhi-110002
Phone: 011-23268786, 23257264
Fax: 011-23281315
E-Mail: synergybooksindia@gmail.com

Acknowledgements

Department of Economics, Jamia Millia Islamia, New Delhi organized a two-day National Seminar on "Economics of Solid Waste Management-An Indian Perspective "during March, 2010 under its UGC Special Assistance Programme in Environmental Economics (DRS-1). The present book is a collection of revised articles presented in this seminar. Our colleagues and students generously blessed us with their help. This help ultimately proved both inspiring and decisive. We are grateful to one and all. We would also like place on record our gratitude to: Professor Shahid Ashraf (Head, Department of Economics, Jamia Millia Islamia) for his constant support and encouragement; All participants in the above referred seminar particularly the contributors to the present volume; Professor Halima Sadia Rizvi, Dr.Indrani Roy Chowdhury and Dr. Bathula Srinivasu (our colleagues); Subanash Prasad Singh (Aurobindo College, University of Delhi); Ajaz Abdullah (Research Fellow, Department of Economics, JMI) and Surjeet Khurana of 'Book Well Publications'.

--M.S. Bhatt and Asheref Illiyan

Contents

Sectoral and Micro Perspective

Contributors

Amita Tayal, Professors, Department of Chemistry, Lingaya University, Faridabad, Haryana

Anurag Srivastava, Centre Head ,Ventura Securities Ltd Agra, UP

Anubhuti Dungdung, Hidayatulla National Law University, Raipur, Chhattisgarh

Arjun K. Singh, CSRD, Jawaharlal Nehru University, New Delhi

Ashfaque Alam, Assistant Director, Association of Municipalities and Development Authorities (AMDA), New Delhi

Babita Aggarwal, Reader, Dept of Economics, MMH College, Ghaziabad, UP

Dimple Vij, Reader, Dept of Economics, AKPG College, Gazhiabad, UP

Ekta Gurjar, Hidayatulla National Law University, Raipur, Chattisgarh

Faisal Zia Siddiqui Asst Professor, AMU & Researcher, Department of Mechanical Engineering, F/o Engineering and Technology, Jamia Millia Islamia, New Delhi

Guarav Tiwari, University of Petroleum and Energy Studies, Dehradun, UP

Krishan K Yadav, Lecturer in Economics, Directorate of Education, Govt. of NCT of Delhi

M Emran Khan, Professor, Dept of Mechanical Engineering ,Jamia Millia Islamia, New Delhi

Mohammed Shakeel, Professor, Department of Civil Engineering, Jamia Millia Islamia, New Delhi

Mrudhusi Swarup, Hidayatullah Natioanal Law University, Raipur, Chattisgarh,

Mohammad Tariq, Programme Coordinator, Toxics Link, New Delhi

Neha, University of Petroleum and Energy Studies, Dehradun, UP

Ravi Agarwal, Director, Toxics Link, New Delhi

Ritika Banerjee, Hidayatullah Natioanal Law University, Raipur, Chattisgarh

Ria Nanda, Professors, School of Applied Sciences and Humanities, Lingaya University, Faridabad.

Rashmi Makkar Panwar, Senior Lecturer, G. B. Pant. Polytechnic, New Delhi- 110 020.

Roopam Srivastava, Assistant Professor, Hindustan College of Science and Technology, Mathura, UP

S.A.M. Pasha, Associate Professor, Department of Political Sciences, Jamia Millia Islamia, New Delhi-110025

Shagufta Jabin, Professors, Department Of Chemistry Lingaya University, Faridabad, Haryana

Sirajuddin Ahmed, Associate Professor, Department of Civil Engineering, Jamia Millia Islamia, New Delhi 110 025. sirajuahmed@rediffmail.com

Sourabh Manuja ,Environmental Consultant, 3/32A, Mendu Compund, Marris Road, Aligarh,U.P

Subash Anand, Reader, Department of Geography, Swami Shraddhanand College, University of Delhi, Delhi

Vishruti Gupta, Delhi School of Economics, University of Delhi, Delhi

Introduction

I

Solid wastes are the most visible forms of environmental bads. Received literature suggests that these are seriously affecting the quality of life and posing serious policy challenges. These challenges are not new but their severity, reach and complex forms are definitely new. Though countries across the globe have put in place policies and instruments but these have invariably proved ineffective. Solid Waste Management (SWM) is a multidimensional and challenging task. In urban areas it has assumed serious dimensions. Due to: lack of appropriate technology, inadequate financial resources, scarcity of land, mixed land uses and absence of effective legal instruments concerned agencies are not in position to deal with the growing magnitude of solid waste. However, it is not correct to suggest that only public bodies are responsible for growing mismanagement of MSW. Polluters are equally responsible. Managing Municipal Solid Waste in a way which protects both environment and human health is a challenging task. With growing urbanization and increased economic growth we have to accord top priority to solid waste management.

Despite policy instruments, administrative mechanisms and legal provisions solid waste management continues to remain one of the most neglected areas of urban planning in India. At present 28% population lives in urban areas in India. Within a few decades this share is expected to cross the halfway mark. This obviously will exacerbate the problems associated with Solid Waste Management further. There has been a significant increase in the solid waste generation in India in the last few decades. This is largely attributed to the rapid population growth and economic development in the country. Per capita solid waste generated daily, in India, ranges from about 100g in small towns to 500g in large towns whereas the average collection efficiency (desired level of efficiency in waste collection) is 72.5%. Around 70% of the cities lack adequate waste transportation facilities. Although there is no national level data for solid waste generation, collection and disposal, data about solid waste generation, are available for a few urban centers. For example, on the one hand the population of Mumbai grew from around 8.2 million in 1981 to 12.3 million in 1991, registering a growth of around 49%. On the other hand, solid waste generation in the city increased from 3200 tones per day to 5355 tones per day in the same period registering a growth of around 67%. If the present situation continues, then the amount of waste generated per capita is estimated to increase at a rate of 1%-1.33% annually. If we take this rate of growth to be 1.33% annually, it is estimated that the total waste quantity generated in 2047 would be approximately above 260 million tones that is, five times the present level. Across India, the production of solid waste also differs in characteristics depending mainly on the size of the population. What is noteworthy is that the biodegradable fraction is quite high in Indian MSW. The ash and fine earth content is also high due to the practice of inclusion of street sweeping, drain silt and construction and demolition debris in MSW.

Solid Waste Management encompasses planning, engineering, organization, administration, financial and legal aspects of activities associated with generation, storage, collection, transfer and transport, processing and disposal of solid wastes in an environmentally compatible manner adopting principles of economy, aesthetics, energy and conservation. It was once commonly thought of as simply "pick up the waste and go dump it in a hole somewhere." Today, nothing could be farther from the truth. When done well, solid waste management successfully blends the diverse interests of a large "stakeholder" community together with industrial interests. Various stakeholders normally involved with solid waste management include: local officials and decision-makers, industry business executives and entrepreneurs, private refuse collectors and disposal site operators, community, neighborhood and environmental organizations, regulatory authorities, recycling service providers, secondary materials processors and end-users. Local governments usually shoulder the responsibility for providing solid waste management services. Laws often empower them exclusive ownership over waste once it has been placed outside a home. Waste management practices differ for developed and developing nations, for urban and rural areas, and for residential and industrial producers. Management for non-hazardous residential and institutional waste in metropolitan areas is usually the responsibility of local government authorities, while management for non-hazardous commercial and industrial waste is usually the responsibility of the generator. Solid waste management has to be area of great concern and interest as unmanaged and ill-treated solid waste poses serious health risks to human beings and the surrounding flora and fauna. All activities in solid waste management involve risk, either to the worker directly involved, or to the nearby resident. Risks occur at every step in the process, from the point where residents handle wastes in the home for collection or recycling, to the point of ultimate disposal. These risks may arise due to the nature of raw wastes, handling of wastes, processing and disposal of wastes. Studies have shown that these risks are higher in developing countries like India as in these countries Solid Waste Management is a labor intensive system. Laborers working in the field are directly exposed to the waste. Even the minimal regulatory framework that exists in most of these countries for environmental protection and occupational health and safety is often not enforced.

We have to generate reliable and up-to-date information and awareness regarding negative effects of solid waste both on human and environment health. Although a number of studies are available on various aspects of waste management but economics of solid waste management has remained a neglected area of theory, policy and research. There is need for an informed understanding of the underlying issues. Received debates and discourses on Economy - Environment linkages have sidetracked this aspect of waste management. Opportunity cost of this neglect is too huge to ignore. There is need for a serious and sustained discourse. With this objective in mind Department of Economics, Jamia Millia Islamia under its UGC sponsored Special Assistance Programme in Environmental Economics (DRS-I) organized a two days 'National Seminar' on: "*Economics of Solid Waste Management: An Indian Perspective*" during March,2010. The main objective of this seminar was to involve all the stakeholders in informed discussions on the various aspects of the Solid Waste Management. Academicians, environmentalists, policy makers, non-governmental organizations, researchers, students and other stakeholders were involved in these deliberations. Articles presented in this seminar and subsequently revised by the authors form the subject matter of the present book.It has been organized in two sections. Section I deals with the macro issues and Section II presents sectoral and regional/ micro perspectives respectively.

I
Macro Perspective

Main focus of this section is to present an overview of the underlying macro issues. Among other things it is expected to provide a proper backdrop for understanding the sector-specific or waste specific problems of solid waste management. As discussed earlier due to rapid population growth coupled with urbanization and economic development. Solid Waste Management (SWM) has become a major environmental and health issue. Ravi Agarwal in Chapter 1 provides an overview of macro dimensions of SWM in India. He argues:"*What this simple concoction of materials, called waste, probably needs is better planning, more public participation, sensible investments and a bucket full of common sense. It is simply not enough to 'sweep' waste out of sight, to make some neighborhoods seem deceptively clean, but it needs a systematic approach involving reduction, collection, safe treatment, recycling and disposal, something which has been beyond the scope of our municipalities".*

Faisal and Emran Khan in Chapter 2 outline current scenario of solid waste in India. They opine that legal frameworks are proving ineffective due to lackadaisical implementation. Their central argument is: *"Although there is an adequate legal framework existing in the country to address solid waste management issues, what is lacking is its implementation. Likely reasons for poor implementation could be a combination of social, technical, institutional and financial issues. Public awareness, political will and public participation is essential for the successful implementation of the legal provisions."*

Formulation of laws and policies are necessary but not a sufficient condition for solving problems. This is true in the case of environmental protection as well. Poor implementation has been a recursive feature of environmental laws. Our constitution has enshrined, in its numerous articles, protection of environment. Over the years courts have interpreted these articles for the protection of environment. For instance, the Article 21 ('Right to life') is a fundamental right. Its scope has been expanded by declaring that the right to life connotes *"quality of life". A* person has a right to enjoy pollution free water and air to benefit from the life completely. Similarly, Article 48A mandates the state to guard and preserve the environment. Guarav Tiwari and Neha in Chapter 3 outline and elaborate these constitutional provisions. The role played by the Public Interest Litigation is also narrated in detail. After tracing major developments in the history of environmental legislation in India they strongly bat for environmental regulation and opine: *"An argument often put forth in developing countries is that it is unfair to ask people to make environmental sacrifices during a period of growth and industrialization when Western countries did not have to make the same choices. This argument does not hold ground in case of India as she faces unique challenges that need unique responses."*

Numerous techniques have been developed for treatment of solid wastes. The state of art technologies developed are:

bio-mechanization, incineration, RDF, composting and vermicompositing. Each of these techniques has merits and demerits. No one has proved effective in India due to a variety of reasons. There is also a deliberate attempt by the Western countries to export outdated technologies such as incinerator technology for conversion of waste into energy. "Waste to Energy Recovery" Plants have not been very successful in India. Options for importing technologies should be explored with great care. These need not be effective by default. Costs and benefits of the available technologies and potential alternatives should be worked out objectively. Two papers one by Shagufta et al (Chapter 4) and second by Faizal and Sourab (Chapter 5) explore the available alternatives and their relative merits. Shagufta et al highlight the importance of Vermicompositing and its relative benefits. They argue:"*Vermicompositing is an easy way to make a positive environmental impact by reducing the amount of wastes. The resulting nutrient rich compost end- product is environmentally sound. This process results in rich manure with high nitrogen (N), Phosphorus (P) and Potassium (K) content and acts as a quality manure in agriculture. In the present context, the Vermicompositing is two dimensional:1) Municipal sludge and other bio-solids are taken into sludge digestion tank where they undergo anaerobic digestion;2) Passage of bio-solids along with green waste into Vermicompositing channel where the organic manure is formed*".

Faisal and Sourab's paper (Chapter 5) explain and compare different MSW treatment technologies using an integrated approach to provide a sustainable solution to MSW problems faced by India. They conclude:"*Wherever it is economically viable and environmentally sound, the option for recycling of waste materials and products should be considered first. The concept of "waste to energy recovery" from MSW in India can be feasible if the waste is properly segregated at source or at the treatment and disposal facility. Since the Indian MSW has a low calorific value, therefore "waste to energy recovery" plants have not been very successful in India. In a landfill, the bio-degradable components of MSW decompose to emit methane – a greenhouse gas 23 times more potent than carbon dioxide and may cause significant environmental problems. Therefore, the bio-degradable component of MSW should be used for energy recovery from landfills*".

Management of waste is the primary responsibility of local self- government institutions such as Panchayats and Municipalities. The manner in which these bodies are supposed to dispose of waste depends on its type and magnitude. Obviously, it cannot be uniform across the board. This speaks for specific sets of rules and guidelines for waste disposal. In view of this formulation of "Municipal Solid Waste Management and Handling Rules 2000" was indeed an important development. These rules require that all cities and towns shall establish suitable MSW treatment and disposal facilities by a certain stipulated time. A significant portion of the Municipal funds (roughly 50%) is to be devoted to waste and sanitation services. Despite these rules the performance of Municipalities in SWM has not been satisfactory. There is a huge gap between performance and promises. Dr Pasha in Chapter 6 delineates the ways and means to reduce this gap. He forcefully surmises:*"Empirical evidence unmistakably tells us that the involvement of women in civic administration in the developing world has brought to the fore positive outcomes. It is necessary to instill the spirit of stakeholdership among population at large. In the fast changing world of science and technology, what we call "waste" is no longer waste, as the latter is converted into energy."*

Among others capacity building is a critical input for waste management. Regular orientation programmes for the benefit of municipal staff can improve their efficiency. Toolkit for preparation of Master Plan for SWM is one of the capacity building programmes through which the municipal officials may be trained to prepare their own sectoral master plan of SWM. Ashfaque's paper (Chapter 7) describes different stages in a tool kit for the preparation of Master Plan for SWM. He asserts that:*"Scientific way of collection, transportation, processing and disposal, financial provisions, public participation and awareness will help in making the city environment neat and clean and will help in long run to reduce the cost of its inhabitants on health/medical treatment. Tool kit for SWM can improve the performance of Municipalities".*

Two sets of instruments viz Command and Control Measures (CCMS) and Market Based Instruments (MBIs) have been employed for the regulation of environmental pollution.

In India both CCMs and MBIs have been put in place from time to time. Due to lack of appropriate implementation mechanism these have not proved as effective as in the developed world particularly in Western Europe. Obviously there is an urgent need for revisiting the environmental regulation and developing appropriate instruments for its regulation. Krishan K Yadav in Chapter 8 justifies such a revisit. The paper first highlights the issues relating to generation of solid waste and then suggests a policy for efficient management of solid waste through pricing of the solid waste generated. He asserts that:" *In order to reduce the quantity of solid waste it is mandatory that less and less quantity is created and even if created, most part of it is recycled at source. Two approaches that can be adopted here are: encouragement and enforcement nature. Encouragement through education and awareness campaigns is essential. However, in case of certain large cities where the problem is getting beyond the tolerance level, enforcement may also be applied in the form of pricing of solid waste created. This price is to be paid by the person generating the garbage to the disposal authorities. Quantity based pricing of the solid waste on one the hand will result in less garbage and on the other hand reduces the financial burden of the disposal authorities".*

II
Sectoral and Micro Perspectives

Eight chapters dealing with sectoral and micro level issues are presented in section II. Economy-environment linkages operate at various levels of aggregation – international, national, meso, and micro level – giving birth to environmental goods and bads with specific characteristics at each level. Macro analysis often sidetracks these specificities. Hence there is need for sectoral and micro level studies. This explains the rationale behind the Section - II. Economists frequently surmise that 'there is no 'free ride' in economics. Everything comes at a cost. Information Revolution has changed the world beyond belief. Unfortunately costs of this revolution seldom receive the desired attention. Policy responses do not even recognize some of these costs. For example with the massive growth of information technology electronic waste (E-waste) is emerging fast as a serious form of pollution. Stocks of 'E-waste' are bound to grow day in and day out. Policy makers are clueless about the potential solutions'[1]. Chapters 9 and 10 by Mrudhusi and Ritika Banerjee and Sirajuddin et al look into the various dimensions of this problem and propose solutions. Mrudhusi and Banerjee focus on the significance of awareness about the various issues involved in generation and management of E-waste. They expostulate: *"Considering the current technological scenario, it is very essential to fathom the importance of issues as grave as e-waste and its management. So, there is a need for the creation of better management techniques and their effective implementation which in turn will generate public awareness. Only when this is in place, will the detrimental effects of e-waste begin to be curbed".*

Sirajuddin et al (Chapter 10) profile the current status of E-waste management practices in Delhi and steps taken by Government and Non-Government organizations to handle it. They argue: *"Solid waste management is a mammoth task in Delhi. Its complications are increasing due to unorganized E-waste disposal particularly computer waste. The phenomenal growth in information technology and outsourcing has given rise to generation of lot of E-waste. Current practices of E-waste management suffer from a number of drawbacks. No inventories are available to quantify waste. Recycling is mostly done by unorganized workers in the informal sector. Exponential growth of E-waste and informal methods of recycling have given rise to toxic materials entering the waste stream".*

Waste generates negative externalities for human health. Festering waste breeds vectors. Mercury from broken thermometers evaporates into the air, and settles back on the water and grass only to come back through milk and fish. All the land dumps leak the toxicity into surface and groundwater, which is never removed before ultimately landing back into our food and water. Industrial effluents, often used as irrigation water for farming, contaminate vegetables and other foods with heavy metals. Numerous international studies have shown a high incidence of respiratory diseases, and chemical body burdens especially in those people who live near waste dumping sites. Thousands of waste pickers live off the discards in our cities, as this is the only job they can find. This seriously compromises their health, through injuries and exposures to hazardous materials. For them it is a non-choice between health and livelihood. Waste is a major contributor to the increasing toxicity of the environment, and reducing this risk has to be the prime objective of any "waste manager"'. (Ravi Agarwal, Chapter 1 of this book). Anubhuti and Ekta in Chapter-11 demonstrate how waste affects health. Similarly, Arjun and Vishruti in Chapter-12 attempted to explore the relationship between poor management of Solid Waste and health by conducting case studies in South Delhi. They plead that: "*There is an urgent need for spreading community level awareness in order to rectify the disposing practices of the households in the study area*".

Based on the field survey Subash in Chapter-13 presents a situational analysis of solid waste in Delhi. He found gross inequality in the distribution of waste collection infrastructure by the Municipal authorities between the posh and the poor colonies. Subash opines, "*Spatial analysis of dustbins, garbage carrying trucks and sanitary workers reveals that their availability gradually declines from core to periphery. Karol Bagh, Sadar Paharganj and Shahdara (S) zones are highly served zones (HSZ) whereas Nazafgarh and Narela are the least Served Zones (LSZ). About 95% Waste is dumped at three landfills and 5% is going for composting. Forty six percent people were not satisfied with MCD services. Around 0.1 million rag pickers, including children and women, are working in extremely unhygienic, unsafe and vulnerable environment*".

General perception is that Municipalities alone are responsible for the management of waste. We cannot put this responsibility on the shoulders of public bodies alone. It is also a responsibility of all residents to maintain, manage and contribute for a better waste management system. People's participation or community participation is very crucial for efficient waste management. It can prove an effective input in minimizing waste and ensuring scientific management. Two papers (Chapters 14 and 15) delineate role and significance of community participation in SWM. In Chapter 14 Tariq showcases decentralized waste strategy in Defence Colony

(A posh colony of South Delhi). Dimple and Babita in Chapter 15 demonstrate how initiatives at individual and community level transformed the waste management concept in the Ghaziabad District of UP. Obviously there is need to scale up such initiatives.

Roopam and Anurag Srivastava in last chapter (Chapter 16) study the existing practices and explore bottlenecks and suggest remedy for SWM in Agra in compliance with the MSW Rules 2000. Their findings assume additional importance in view of Taj Mahal being an international tourist destination.

Endnote

1. Bhatt M.S, Ashraf Shahid and Illiyan Asheref (2008),"Problems and Prospects of Enviroment Policy-Indian Perspective", Aakar Books, New Delhi, pp 22

Chapter 1

Waste – The Business of Exclusions

Ravi Agarwal

Waste is exploding in India. As more and more people live in cities (expected 340 million by 2010), and consumption increases, waste streams too are growing. The major metros are expanding rapidly, while previously small towns are burgeoning into small cities. What was two decades back, mostly food waste, now contains electronic waste, construction waste, all types of plastics, metals, medical waste and also industrial waste mixed with it all, like a toxic cocktail. Most of all this still lands up in dumps often-situated in low-lying areas, which were previously water bodies. Pigs, cows and waste pickers rummage through waste piles on street corners, and passer byes hold their noses as they sidestep the decaying morass. Post monsoon, most of our green cities are subsumed by fallen leaves and cut grass, which is then piled up and burnt, the sooty smoke descending all around. Household hazardous waste also abounds. Batteries, old pesticide cans, mercury thermometers, used syringes, florescent lamps all find their way into the waste bin. On the other end, small and tiny recycling operations, hidden behind high brick walls, melt the plastic, rubber and metal down, emitting noxious pollutants into the air. At other places old computers are broken down, their parts melted in strong acids, or burnt to recover valuable materials. At best here, waste is 'dirty' business! Simultaneously the powerful international waste industry is offering state-of-the art waste processing facilities, provided it has access to the waste, both from here and from imports, even though there have been only a string of technology failures so far. Municipalities have meanwhile failed to implement even simple systems, leave alone complicated ones. While expert consultants and the promise of new funds (JNURM, the Twelfth Finance Commission, and CDM), are seducing city governments with high tech investments, and sophisticated jargon, what this simple concoction of materials, called waste, probably needs is better planning, more public participation, sensible investments and a bucket full of common sense. It is simply not enough to 'sweep' waste out of sight, to make some neighborhoods seem deceptively clean, but it needs a systematic approach involving reduction, collection, safe treatment, recycling and disposal, something which has been beyond the scope of our municipalities.

Waste has a significant consequence, especially on health. Waste becomes a concentrated repository of all that is undesirable. Festering waste breeds vectors and old discarded tires can hide water, which breeds mosquitoes. It can contain very toxic compounds and heavy metals, and can produce greenhouse gases like methane, and very toxic gases like dioxins when burnt. Mercury from broken thermometers evaporates into the air, and settles back on the water and grass only to come back through milk and fish. In the US alone over 40% of the lead in landfills is from electronic waste. All the land dumps leak the toxicity into surface and groundwater, which is never removed before ultimately landing back into our food and water. Industrial effluents often used as irrigation water for urban farming, contaminate vegetables and other foods with heavy metals. Numerous internationally studies have shown a high incidence of respiratory diseases, and chemical body burdens especially in those people who live near waste dumping sites. Several hundred thousand waste pickers live off the discards in our cities, often since this is the only job they can find, and seriously compromise their health, through injuries and exposures to hazardous materials by doing so. For them it is a non-choice between health and livelihood. Waste is a major contributor to the increasing toxicity of the environment, and reducing this risk has to be the prime motive of any 'waste manager.'

Currently we generate over 45 mt of municipal waste each year, which is expected to cross 65 mt by 2010. Of this over 15% is plastic waste, while a major portion of the volume is construction malba. E waste is at 3.3 mt. while hazardous waste pegs at 7-8 million tonnes. However the waste generation is not constant across economic classes. Easily the rich are responsible to the most generation. A household, which earn about Rs. 3000 per month would generate less than 150 gm of kitchen waste per day, while one, which earn over 10,000 per month, could generate over 800 gms per day. Consider also that computer density of about 65 per thousand will be in the middle and rich classes, and that over 2 billion pet bottles are used by consumers of mineral water and soft drinks, the class character of waste generation becomes evident. Not only those who earn more generate more waste, but also the industrial activity, which benefits them most, contributes to waste generation. On the other hand, landfills are sited on the cheapest land available, and so is the location of recycling and waste treatment facilities. The poor obviously generate the least amount of waste, but feel its impacts the most.

Waste as trade

It is not by chance that India is a major recipient of waste internationally. Labour is cheap, and recovery of materials from waste relatively inexpensive. Plastics, electronics, even medical waste has been found to be imported into India, legally or illegally. Customs documents are misclassified to hoodwink port authorities and their port of origin fabricated to make them impossible to track. Several studies and court submissions have revealed the tip of the iceberg of what is probably a massive activity. In specific instances waste oil, asbestos laden steel, lead waste has been stopped by court orders. In Export Promotion Zones (EPZ) waste is imported, reprocessed and the materials recovered set back. However what remains is the waste or hazard, which no one wants and remains on our shores. Recent conservative estimates peg e - waste imports to exceed 50,000 mt annually. The dynamics of cheap labour and easy entry also guide the logic of India being a major shipbreaking country in the world. In fact these activities are justified internationally on the basis of an offer' cheap' labor, even though at home little is done to protect labour conditions. It is now history that India, along with the G 77 countries was one of the key nations demanding that waste dumping be stopped from developed to developing countries under the Basel Convention. Today, the situation is reversed as India openly argues for waste imports and sees this as an opportunity to be a waste processing country globally even though locally produced waste is left uncollected and continues to pollute local environments.

Business not health

In all this it is evident that the language around 'waste' has been transformed in the recent past. Waste is now promoted as a business first, in internationally sponsored workshops and seminars, even though such business hides a host of public subsidies. Waste as wealth, is considered not in the framework of dealing with waste as a public health hazard, but falsely as an economic good, even though even in developed countries, public money subsidizes waste systems. The waste industry is a mammoth one globally, exceeding an estimated over 6 billion USD in the US alone. In Italy the mafia controls waste, as the recent waste crisis in Naples amply demonstrated. In India too the waste industry is on the threshold of stepping in. Major investments have been made in e waste recycling units, hazardous waste units, and medical waste disposal units. While infrastructure development is key to proper waste management, there is also the danger of creating a powerful industry, which needs waste to survive, even as on the ground systems like local collection and segregation continue to fail or are insufficiently focused upon. For such an industry, more waste is good news, and the ideas of waste minimization, waste reduction, waste reuse, and design for longevity are 'dirty' words. Alongside, such industry is also rooting for the opening up of the international waste trade, since large investments need continuous raw materials for them to be profitable. It is this danger, which can and needs to be avoided. Sophisticated systems can only be built upon simple successes.

Natural resources

Waste has a direct connection to the consumption of raw materials and energy and arguably it represents inefficient transformation of minerals and raw materials into products. The life cycle of all the products we use lay bare a strew of waste and toxics in their trail of being mined as mineral, transformed into usable materials like steel, plastics and aluminum and then being discarded back as waste. If this loop is not closed down it essentially implies that we are constantly converting nature into waste. Hence all good and sustainable waste practices look at life cycle impacts and 'close the loop.' Policy instruments like extended producer responsibility bring in the idea of end-of-life of products back into the production life cycle, and ensure that safe disposal is part of product and systems design. Hence a computer sold in Sweden already has a cost for recycling it built into it, and the industry association which collects the used computer in partnership with the local municipality, need to ensure that it goes to an authorized recycling facility. Such looping back has also changed the manner in which, for example, automobiles, which took over large land mass when junked, are being designed in Europe, to make them easy to dismantle and recycle.

Alongside waste disposal uses up a lot of land and creates conflicts over land use. In Delhi, the municipality has been fighting to convert a portion of the protected local ridge forest, a green lung of the city, into a landfill, instead of planning to reduce its growth. Hence sustainability of an efficient economy must build into it the idea of waste and preventive aspects of reducing impacts have to be seriously considered.

Legislation and Judiciary

The past decade has seen a flurry of legislations dealing with waste in India. However some have been more effective than others. Beginning from the Hazardous Waste Rules in 1989 (last amended in 2003), the Bio Medical Waste Rules (1998), a law to deal with Lead Acid Batteries Rules (2001), the Municipal Waste Rules (2000), the Plastics Waste Rules (2011), to the most recent guideline to deal with E - Waste Guideline (2008), these are based on different waste streams. Some like the Hazardous Waste Rules have triggered off public private partnerships and currently 22 facilities to deal with industrial waste have been set up in the country. Similarly, the Bio Medical Rules have resulted in the setting up of 160 facilities around the country for collection of medical waste from hospitals and clinics, based on user fee. This particular legislation also mandates a ban on the incineration of PVC plastics, which is globally recognized as a dioxin generator. However across the board quality of these services are matters of great concern. Typically, despite the Rules and Guidelines medical waste is burnt in illegally incinerators in centralized facilities, even though this is severely restricted by law and the safer way of autoclaving is not adopted. Similarly a recent major fire at the Ankeleshwar facility for treating hazardous waste, and promoted as one of the best, revealed that while the facility should have stored only 60 tonnes of waste more than 6000 mt was on site. This is despite national standards on such operations which include landfill design and incinerator emission needs. Clearly, even though private actors are investing in infrastructure, regulation of operations and emissions is wanting and quality of disposal is dismal, with attention to detail lacking.

The Courts have played a major role in the past decade. The first ban on the import of hazardous waste into India was set up by the Delhi High Court (Srishti, April, 1996) and later fortified by the Supreme Court (SC, May 1997, RFTE vs. UOI)). It still holds. The SC also set up the MGK Menon Committee for a plan to manage hazardous waste in the country, which gave its far-reaching report in 2001. The Bio Medical Rules were initiated after an NGO Srishti intervened

(1996) in an ongoing case (B.L.Wadhera vs. Union of India), and were promulgated in 1998. The Municipal Waste Rules came along with the prodding of the SC (Almitra Patel vs. UOI), though under the same case where the petitioner was arguing for better landfills in each city, the Court also passed strict orders to remove slums. Similarly there have been several cases in High Courts and local courts, dealing with a range of issues like plastics, landfills, waste to energy etc. The judiciary has hardly been a silent spectator, and has led from the front.

A general fear of being overrun by regulation is resulting in the Government being reluctant to legislate even where it may be essential. The recently issued E - Waste Guidelines are a case in point. Only very recently (2010) has a draft law been published, but it is still awaiting notification. Hence even though e waste is one of the fastest growing waste streams both in India and abroad, India still does not have a requirement for its disposal. Most of this waste is dumped or recycled in backyard operations, which attempt to recover precious materials like gold and copper from the waste, rather than recycle it as a whole. To avoid such cherry picking demands a regulation where the objectives are spelt out and is not merely a voluntary guideline. The Industry too, which represents the heavy weights of Indian business is looking for a level playing field.

On the other hand a legislation made to trigger producer responsibility and take back systems in lead acid batteries has not succeeded. The battery industry has not been able to develop adequate collection systems even though it had been mandated to do so. Hence a car battery exchanged with the dealer for a new one often land up in highly polluting backyard operations. Of course any such legislation also needs detailed working on the ground, and there can be no short cut to do this. This legislation is also under review at the moment.

Industry's reluctance to take real responsibility for waste has not helped matters. In the case of plastics, for example, a high level committee headed by Justice (retd.) Ranganathan

Mishra (2001) received an assurance from the Plastic Manufacturers Association that they will set up plastic PET bottle collection centers throughout the country and also provide a 25 paisa bottle refund for those bottles, which are brought to such centers. Over a decade later, no (or hardly any) such facilities exist. Ironically, the Association has been promoting school programs to educate school children why plastics are 'essential' to life! Of the over 4 million tonnes of plastics packaging manufactures, over 50% end up as waste soon after usage. Plastic bags have been recovered from cows stomachs, bird innards and are ubiquitous everywhere. There is widespread concern in the public about them, and even the rural countryside is replete with them. Many States like Delhi and Haryana have attempted to ban plastic bags, only to be thwarted at the last moment by some unseen hand. Instead a national law, which bans bags of 40-micron thickness, has been now promulgated. However though this has ensured that only carry bags made of virgin plastic are available, it has not helped in the litter being reduced. Very recently, a new law (Plastic Waste Management and Handling Rules 2011) has been notified. This mandates the setting up of collection systems in each municipality, along with industry responsibility, charging for carry bags, tracking of waste as well as a ban on small multilayered plastic packaging.

Community Efforts

Over the past decade, there has been a spontaneous response from citizen's to improve cleanliness in cities. These are largely small scale and local, but represent a common desire to live in clean surroundings. NGOs have helped with their expertise, and also attempted to link the issue of livelihoods through involving waste pickers in door-to-door collection. Exnora in Chennai, Toxics Link and Vatavaran in Delhi, Stree Mukti Sangathan in Mumbai, NEEDs in Shillong, NBJK in Bihar, CDC in Rajasthan are some of the more organized efforts. However there has been a response from Resident Welfare Associations, Housing Cooperatives, and Institutions like the IITs to set up local systems. In many places city officials like municipal commissioners have played an exemplary role. The problem of waste and sanitation in low-income colonies or in shantytowns is more severe, even though many NGOs have successfully intervened and created better systems on the ground. The municipality though extends partial or no services to these areas. There is also little incentive for them to create infrastructure on the ground since landuse is disputed, and displacement of such settlements has become common across cities, as land prices have soured.

It is in such settlements that the waste pickers reside as well. Hutments are packed with collected paper and plastic to be sorted and sold to the local kabari. Often in the heat of summer, whole basti's have caught fire, fuelled by the waste. In a major fire in Jahangirpuri in Delhi in 1995, the plumes were seen for miles around and it was believed that the smoke was very toxic. Such incidents do not lead to better conditions within these bastis, but become another rationale for shifting them to the new margins of the city. In this way the slogan of 'Clean and Green' has led to more injustice in the name of environmental betterment. It is thus no surprise that while the clean and green agenda is supported by the rising middle class; it is paid for in human cost by those marginalized by income and caste.

Unfortunately, even in places where there have been successful community based interventions, whether in affluent or in poor colonies, there has been little attempt by the Government to help upscale these efforts. In a significant study carried out by Toxics Link to determine barriers to up scalability of successful community based projects nationally, it was clear that there was almost no real support available to help such initiatives. They are looked upon rather patronizingly, as the efforts of good citizens, rather than as viable models of city waste management. In fact even land for local composting is not provided. Typically a local intervention would service 4 - 5000 households, though this could extend to 100,000 households, as in the case of Muskan Jyoti in Lucknow. These NGO and community led intervention involved waste pickers in door-to-door collection and local colony level composting. These models also augment the waste pickers income by safeguarding their access to recyclables. They have provided dignity and security to them by providing them recognition through ID cards, and made them less vulnerable to constant harassment by the police. They are good examples of integrating livelihoods and environmental sustainability. However large-scale privatization of waste collection, as is currently the trend, is only leading to a larger marginalization. It seems the state is more comfortable dealing with a single private player rather than a range of community actors.

Informal Sector

The informal sector has been collecting recyclable waste for decades in cities in India. They are mostly low skilled operators, who develop the fine art of distinguishing between different types of plastics, metals and essentially what will fetch them a higher price with the local 'kabadi.' Some have fixed beats and protect their own waste bin, while others roam the streets scavenging what they can find. On a landfill site the hunt for materials become even more nuanced. Every nail is picked up using ingenious methods like magnets on long poles, or dirty plastic bags, which no one lese will touch are cleaned and washed. Clearly such detailed rummaging is something, which no machine can hope to achieve. However this livelihood is fraught with risk. Injuries, cuts and bruises are common. Exposure to infectious waste, mercury, lead and other dangerous chemicals is unavoidable. Many of the waste pickers are women and children. For them any livelihood is better than no livelihood.

However, despite many policy averments, to support waste pickers, including in the National Environmental Policy (2006), and earlier in the Supreme Court's Burman Committee (1999) and the Planning Commission's Bajaj Committee (1995), there has little investment by the state in their future. In fact the recent trend, especially under the urban renewal missions, to subcontract waste collection to private operators by municipalities has worsened the situation. Many of these private operators, deny access to the waste pickers to the bins they have been collecting from for decades. If there is money to be made, the private operators rather make it themselves than allow the booty form recyclable waste be shared. Hence some operators have begun the side business of collecting and selling recyclable waste, depriving the waste picker of his or her livelihood. Such mindless centralization is harmful, besides it is more costly. Private parties link to other technology investors in recycling and waste processing or waste landfills, and all waste goes into centralized places. Spaces too are not provided, and in fact in the new Master Plan of Delhi 2021, waste dealing (*kabaris*) or waste separation activities are not a recognized, even though they form part of colony infrastructures. While the city may seem cleaner, it is not more sustainable since waste, which could be recovered, is dumped and local communities have no participation in such systems.

Many NGOs and their collectives have been raising this issue trying to protect the waste picker from being excluded. A new initiative led by SWaCH (Pune) is bringing these groups (such as SWM - Mumbai, KKKP - Pune, Chintan and HRA - Delhi), together and they are proposing a new policy to protect livelihoods. What is needed is a mix of local and central systems, which allow for optimum collection and recovery of materials, and have strong ground level systems.

Waste to energy

Waste to energy has been promoted as the future of waste in India. As early as 1995, when the Ministry of Non Conventional Energy announced the policy to subsidize any energy produced by waste, it has been packaged as a win-win. However of the over 90 projects implemented or planned so far, hardly any has succeeded. Technologies like bio - methanation, incineration, RDF, even composting and vermi-composting have been tried, but non have much to show for success. The reasons for failure are many, but could largely be owing to inappropriate technologies as well as poor systemic linkages. Take the case of incineration of waste to recover energy. The earliest plant installed was in Timarpur in Delhi, in 1984, where a Danish incinerator set up for a cost of over 34 crore rupees ran for less than a week. There had been severe miscalculations on the density of the waste and its calorific values, in spite of top technologists who did the equations. As late as 2006, a well-accepted technology based on bio-methanation was installed by a private company in Lucknow, with the blessings of the local MP who was also India's Prime Minister, and subsidy by the World Bank. The plant today is closed, and has been subject to Supreme Court investigations about the reasons for its closure. While waste in Europe has almost 50 percent of the materials as plastics and paper, Indian waste is wet, food, and with little plastic or paper. Hence if incinerated more energy is put in than taken out! Where bio - methanation has been used, which converts waste into methane and then energy, municipalities have suddenly become very possessive about their waste, since they would rather sell it than just give it. If there is money to be made, why should they not too partake? Thirdly the energy produced is more expensive than that sold by subsidized State Electricity Boards, who are not willing to buy it. All these reasons also hide a subsidy- in- perpetuity scenario.

The question of safe technologies is an important one per se. Globally, incineration of waste has been severely regulated or outright banned (as in the Philippines) since the emissions of very toxic compounds like dioxins and heavy metals like lead and mercury push control technology cost up exponentially. A 2000-ton per day incinerator in Europe can cost upwards of USD 500 million and easily one of the most capital-intensive projects a city can undertake. Such high cost investments then scout for more waste to feed the guzzlers, and it is not unusual for international waste processing companies to scout for waste world wide. The energy from them is not necessarily cheaper than from other sources, and has to be weighed in against the comparative cost of recycling. Globally large civil society networks such as GAIA (Global Alliance for Incinerator Alternatives) resist such investments both on the logic of pollution but also their detrimental impacts on local waste systems of composting and recycling. In India there are 30 such proposal for waste to energy processing plants, which have mushroomed after the recent access to climate change based carbon credits, and JNURM based funds.

Besides siting of such projects is a major challenge. Local protests have recently greeted one such project, which is proposed as an Integrated Waste Facility, based on a contested technology called RDF, (which converts waste into concentrated fuel pellets) to be setup close to a large colony in East Delhi. There is little information made available to the local residents, nor their consent sought about the siting. The community has serious concerns; amongst them is the burning of PVC based plastics, which is illegal in India.

New Funds

Till recently all waste management was solely funded by municipal tax collections. A significant portion, upto 50% of this money was devoted to waste and sanitation services. Over 30 % (upto 50%) of the municipal staff has been employed towards this end. The situation has now changed for larger projects. The Central Government has under the Twelfth Finance Commission allocated over Rs. 2500 crores for solid waste management. Over Rs. 10,000 crores can be availed by 63 cities under the JNURM for this use before 2010. Subsidies are available for composting and waste to energy. State governments are providing subsidized land, and various types of tax holidays as well for infrastructure development. Carbon Credit and Carbon Finance are available for eligible projects under the Kyoto Protocol. However there needs to be a more sustainable mechanism for the operations of these ventures, and the waste generator ultimately has to pay through a 'tipping' fee, paid by the waste generator. This issue is a political hot potato, and unless municipalities bite the bullet, waste infrastructure projects are not likely to succeed for a very long time. Also it is clear that while large capital-intensive projects are being supported by state funds, the same are not available to community based low investment projects.

Future Systems

Would it not be wonderful if everything we use were

totally recyclable or reusable? That the materials we use are totally non-toxic, and do not people or the environment. Waste is a 'negative' good, undesirable and which society bears a cost towards. Often the cost is hidden in terms of the health and environmental impacts, but accounting for them helps in appropriating them where they must be borne. This is not a dream but the pillar of what the future of waste management is being built upon. There are two requirements for this. Firstly that there is adequate infrastructure to recycle whatever becomes waste, and re-introduce these materials back into the economy as raw materials. For example if the computers we use can be broken down and their individual components broken down to basic materials like plastic, glass, metal completely then these can be made the new parts afresh. Secondly it requires that all the hazardous components be removed from all the materials we use. Hence there should be no lead in the displays, no mercury in the switches, no lead in the solder, or no flame-retardants in the plastic, or no PVC in the electrical wires. This would make recycling much simpler and much safer. Both of these need fundamental changes, some of which are already happening through new laws like the WEEE and ROHS in Europe and other countries.

Along with such basic systems guided by principles of sustainability, new systems need to recognize and support local waste management and resource recovery. Both upstream 'preventive' as well as downstream (improved management and handling) approaches are needed together. Indian waste is largely organic, and as per World Bank figures is likely to remain so for the next three decades. In such a scenario, compost along with energy products makes eminent sense. With local composting comes also job creations and an opportunity to involve existing waste pickers in door-to-door collection schemes. Even in hazardous waste streams like that of e- waste, which need larger and more sophisticated facilities to process the waste, the informal sector can play the key role of a collection agency. In this manner, they still use their skills to earn a livelihood with the hazardous component of the work removed from their ambit. The private sector has an important role, since it brings in a new efficiency compared to municipalities, but it needs to sorely dovetail with existing systems not compete with them. It needs to do more than use waste bins as ad-posts!

Like in many other sectors, India is opting for systems, which are very capital intensive and centralized. This is since most of the dominant players in the waste business operate in countries not confronted by the challenges of livelihoods and environment being combined. In India, by blindly following the logic of the west, we will end up creating greater injustice, social tensions and exclusions. On the other hand build up from what is there, instead of replacing it, and learning from the best the world has to offer, will utilize the systems present here but which the more advanced 'west' is now trying to revert to but cannot. Our cities of the future cannot be islands of prosperity, if the people in them are not participating in a just and equitable manner. How we think and deal with waste is a key test of sustainable development in the rapidly urbanizing India, a challenge not faced by any other country in the world today. The way we do it, will impact the globe for a very long time.

Chapter 2

Solid Waste Management in India

Faisal Zia Siddiqui and Khan M Emran

Introduction

The Ministry of Urban Development (MoUD) have assessed the solid waste generation in India to be about 0.1 million metric tonnes per day (MM/d) in the year 2001-02. The Central Pollution Control Board (CPCB) made a survey of 59 cities in India (2004-05) to assess the existing status of solid waste management which included 35 metro cities and 24 State capitals. Based on this study and on census data of 2008, the solid waste generation has been estimated to be 0.573 (MMT/d) in the year 2008.

There are about 4377 municipalities and municipal corporations spread throughout the country. The average collection efficiency of solid waste ranges from 22% to 60%. The highest per capita waste generation is in the city of Kochi (0.67 kg/capita/day) and the lowest is (0.17-0.19 kg/c/day) in Kohima, Imphal and Nashik. The waste characterization shows that it typically contains 51 % of organic waste, 17%

recyclables, 11% hazardous and 21% inert. However, about 40% of all solid waste is not collected.

Regulatory Framework for MSW Management

The MSW (Management and Handling) Rules, 2000 was notified by Ministry of Environment and Forest(MoEF) for management of Municipal Solid Waste Management (MSW). The main objectives of MSW rules are to provide scientific management of MSW; to ensure proper collection, segregation, storage, transportation, processing & disposal of MSW; to upgrade existing facilities to arrest contamination of soil and ground water.

The MSW Rules apply equally to every municipal authority regardless of its size. The rules provide directions for segregation, collection, storage, transportation, processing and disposal of MSW. Also the rules require that all cities and towns establish suitable treatment and disposal facilities by a certain stipulated time. The Rules specify the landfill site selection criteria and standards for composting, leachate control and management and closure of landfill sites. The 4R's of waste management i.e. waste reduction; reuse, recycling and recovery are encouraged.

Status of Plastic Waste Management

Plastics waste has attracted widespread attention in India. The plastic consumption in India, as per estimate in 2008 by CPCB (Central Pollution Control Board) was 8 MT/annum, out of which about 5.7 MT of plastics are converted into waste annually i.e. 15,722 tons of plastic waste is generated per day. Therefore the per capita generation of plastic waste has been estimated as 5.7 Kg/annum. The figures available on plastic waste are estimated on the assumption that 70% of the total plastic consumed is transformed into waste. It has been reported that 60% of total plastic waste generated is recycled and 40% is littered and remains uncollected. Therefore, approximately, 6289 tons per day (TPD) i.e. 40% of plastics are neither collected, nor recycled. The plastics waste may be Thermoplastics which are recyclable, or Thermosets which are not easily recyclable.

Thermoplastics, which include Polyethylene Terephthalate (PET), Low Density Poly Ethylene (LDPE), Poly Vinyl Chloride (PVC), High Density Poly Ethylene (HDPE), Polypropylene (PP), Polystyrene (PS) etc, constitute 80% of the total plastics. Thermoset plastics comprise alkyl, epoxy, ester, melamine formaldehyde, phenol formaldehyde, silicone, urea formaldehyde, polyurethane, etc. which constitute the rest 20% of the total plastics waste generated in India. Commonly littered plastic wastes include mainly the polythene carry bags, plastic wrappings, thermocol packing, plastic plates, cups, spoons etc. as recyclable plastics and gutkha /supari pouches, multilayer packaging, laminated packing etc. as non-recyclable plastic waste. In terms of types of plastics, almost 90% of the plastic types are recyclable and only 10% of the various types of waste are non-recyclable. Thus, recycling of plastics reduces the requirement of most of the plastic.

Regulatory Framework for Plastic Waste Mangement

India notified the Recycled Plastics Manufacture and Usage Rules, 1999 (as amended in 2003) prohibiting the use of recycled plastic bags for storing, carrying, dispensing or packaging foodstuff. These rules also provide that minimum thickness of bags should be 20 microns and the size should not be less than 8" x 12". The prescribed authorities for use, collection, segregation, transportation and disposal under these Rules are District Collectors (DC) and Deputy Commissioners (DCs) for plastic carry bags and the State Pollution Control Board (SPCBs) and Pollution Control Committee (PCCs) for manufacturing and recycling of plastic waste.

Status of Bio-Medical Waste Management

Approximately 288.20 TPD (56.87%) out of 506.74 TPD wastes generated is being treated either through Common Bio-medical Waste treatment facilities (159 in number), or captive treatment facilities (CTF). There are 602 Bio-medical Waste Incinerators (which include both common and captive incinerators), 2218 autoclaves, 192 microwaves, 151 hydroclaves and 8,038 shredders in the country. About 424 (70.4%) out of 602 incinerators are provided with air pollution control devices.

Regulatory Framework for Bio-medical Waste Management

The MoEF notified the Bio-medical Waste (Management & Handling) Rules (BMWM Rules), in year 1998 under the Environment (Protection) Act, 1986 for safe handling, segregation, storage, transportation, treatment and disposal of bio-medical waste generated from the Health Care Establishments (HCEs)/Health Care Facilities (HCFs). These Rules were amended in 2000 and 2003 so as to fill the gaps as felt necessary at that time. These Rules are comprehensive and have stipulated duties for every bio-medical waste generator, occupier, transporter and CTF operator to maintain source separated waste, pack, store, treat and dispose these wastes as per the laid down provisions. The SPCBs/PCCs have been notified as the prescribed authority for implementing the provisions of these Rules in the respective States/Union Terrirories (UTs).

Status of E-waste Management in India

The increase of electrical and electronic products consumption rates and higher obsolescence rate leads to higher generation of e-waste. The increasing obsolescence rate of electronic products also adds to the huge import of used electronics products. Electrical and electronic waste (E-waste) is one of the fastest growing waste streams in the world. The increasing "market penetration" in developing countries, "replacement market" in developed countries and "high obsolescence rate" make e-waste one of the fastest growing waste streams in the world. The E-waste inventory based on the obsolescence rate in India for the year 2005 has been estimated to be 1,46,000 tonnes, which is expected to exceed 8,00,000 tonnes by 2012. Dumping of E-waste poses a major challenge in India. Imports are one of the major sources of E-waste in the country. The accurate data on such imports is not available largely owing to the nature of the trade.

Regulatory Framework for E-waste Management

There are no specific environmental laws for E-waste In India. E-waste has been included in the Schedule IV of the Hazardous Waste (Management, Handling & Transboundary Movement) Rules, 2008. This inclusion makes it mandatory for any E-waste recycler to get it registered with Central Pollution Control Board (CPCB). Recognising the fact that E-waste is seen as a major future problem, guidelines for "Environmentally Sound Management of E-waste" has been brought out suggesting treatment options and adoption of environmentally sound E-waste treatment technologies. E-waste has been categorised in Schedule 4 of the Hazardous Waste (Management, Handling and Transboundary Movement) Rules, 2008 under the "Components of waste electrical and electronic assemblies comprising accumulators and other batteries included on list A; mercury-switches, activated glass cullets from cathode-ray tubes and other activated glass and PCB-capacitors, or any other component contaminated with Schedule 2 constituents (e.g. cadmium, mercury, lead, polychlorinated biphenyl) to an extent that they exhibit hazard characteristics indicated in part C of this Schedule". These can now be recycled by the units registered with CPCB and having Environmentally Sound Technologies. The Extended Producer Responsibility (EPR) concept had been introduced in the guidelines. All SPCBs have been requested to grant consent to establish and give authorizations for recycling of Waste Electrical and Electronic assemblies, so as to enable the units to register as recycler with the CPCB.

Status of Hazardous Waste Management in India

There are about 36,000 hazardous waste generating industries in India which generate 6.2 million tonnes out of which land fillable hazardous waste is about 2.7 million tonnes (44%), incinerable hazardous waste is about 0.4 million tonnes (7%) and recyclable hazardous waste is about 3.1 million tonnes (49%). There are 141 hazardous waste dumpsites that have been primarily identified in 14 States/UTs out of which 88 critically polluted locations are currently identified. Gujarat (about 29%), Maharashtra (about 25%) and Andhra Pradesh (about 9%) are the top three Hazardous Waste (HW) generating States. Thereafter, Chhattisgarh (about 5%), Rajasthan, West Bengal and Tamil Nadu (about 4%) are found to be major generators of HW. These seven States together, are generating about 80% of country's total HW. About 64 Common Hazardous Waste Transportation, Storage and Disposal Sites (TSDFs) have been identified in various States/ UTs out of which 35 sites have been notified. 25 TSDFs are operational and 9 TSDFs are under construction. As per the National Inventory of Hazardous Waste Generating Industries, total waste handling capacities of TSDFs is about 1.5 Million Tones per Annum (MTA) and there is a deficit of about 1.2 MTA for land fillable wastes and about 0.9 MTA for incinerable wastes.

Regulatory Framework for Hazardous Waste Management

Ministry of Environment and Forests has notified the Hazardous Waste (Management, Handling and Transboundary movement) Rules, 2008 repealing earlier rules so as to make the Rules more practical and effective in implementation. The National Hazardous Waste Management Strategy is being finalized for effective management of hazardous wastes so as to avoid environmental pollution and adverse effects due to improper handling and disposal. A Coordination mechanism for effective implementation of Hazardous Waste (Management, Handling and Transboundary Movement) Rules, 2008 by the custom authorities at the ports has been established.

Key Stakeholders in the Solid Waste Disposal Sector

There are a number of government organizations who play a key role in India's solid waste disposal sector (Refer figure-1) such as local municipalities/urban local bodies, Ministry of Environment and Forests (MoEF), Ministry of New and Renewable Energy Sources (MNRE) and Ministry of Urban Development (MoUD). In India local municipalities have the overall responsibility for the Solid Waste management. They implement laws created by MoEF. The various government organizations and stakeholders involved in the management of Solid Waste are:

- **MoEF:** MoEF brought out the "MSW Management and Handling Rules 2000". It governs the CPCB, which is responsible for the oversight of the 2000 Rules. The MoEF is responsible for general aspects related to waste management, coordinates the environmental policy and overseas the implementation of the legislation regarding waste management.
- **MoUD:** MoUD is a nodal Ministry overseeing urban development in India. The Ministry's Central Public Health and Environmental Engineering Organisation (CPHEEO) guides the states and municipalities on technical aspects of solid waste management. As such, CPHEEO is conducting data collection on methane extraction potential from landfills in the country.
- **MNRE:** MNRE covers both renewable energy and new fossil fuel technologies. The Ministry has recently indicated that it is willing to consider making subsidies available to Land Fill Gas (LFG) projects.
- **CPCB:** Coordinate the activities of the SPCBs and provide technical assistance and training to the personnel. Disseminate information sponsor research relating to waste management. **SPCBs:** plans a comprehensive programme for the prevention, control or abatement of air pollution and water pollution. To inspect, any control equipment or process ,prior to installation of a landfill or incinerator, permission from SPCB must be obtained.
- **Municipalities/Urban Local Bodeis(ULBs):** They provide waste management services or operates disposal, recycling, or composting facilities. Often contracts out services to the private sector.
- **Private Formal Sector:** Private Formal Sector participates in performing the functions of collection and transportation of the waste and may operate disposal and recycling facilities.

- **Private Informal Sector/Rag Pickers:** Private Informal Sector participates in collection of the recyclable waste, transfer to the recycling facility and recycling of waste.
- **Service Users:** Service users are the people who use the sernmunicipalities/ULBs.

Figure 1: Role of Various Bodies in Waste Disposal Sector

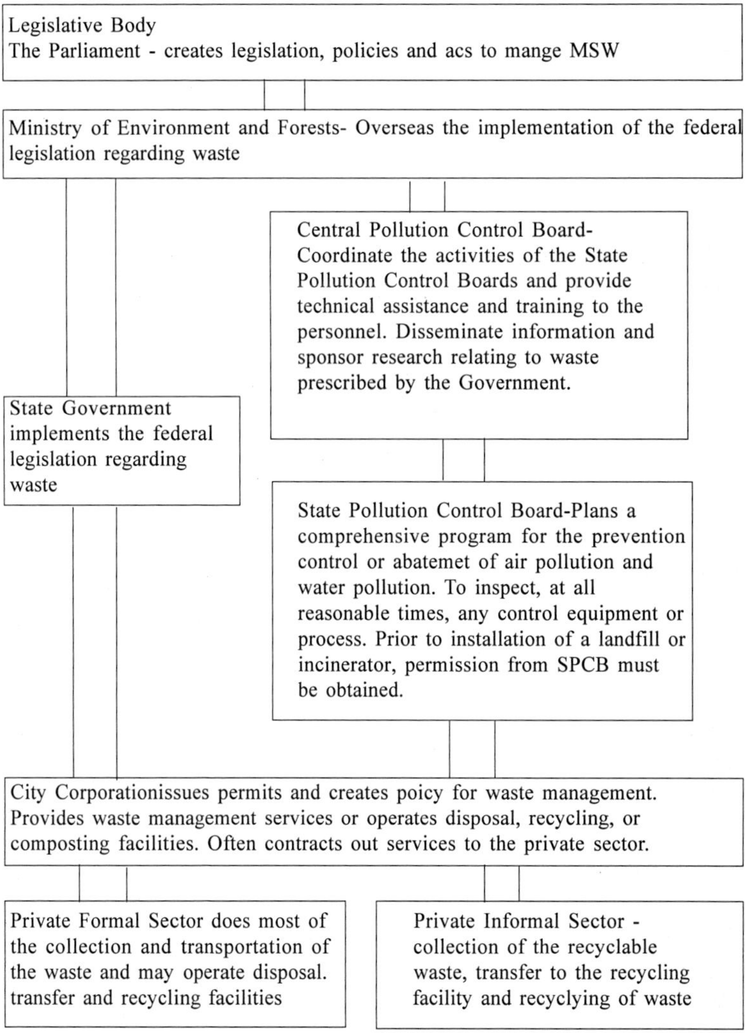

Conclusion

The rapid increase in the population is changing the nature of solid waste management from a low concern, localized issue into a global social problem. Risks to the public health and the environment due to solid waste in large metropolitan areas are becoming unbearable. The paper has summarized the salient features of the current scenario of solid waste management in India. India currently is facing a solid waste dilemma, for which all elements of the society are responsible. The community sensitization and public awareness is low. There is no system of segregation of solid wastes at household level. Although there is an adequate legal framework existing in the country to address solid waste managemnet issues, what is lacking is its implementation. In spite of a stringent legislation in place, open dumping is the most wide spread form of waste disposal. The likely reasons for poor implementation could be a combination of social, technical, institutional and financial issues. Public awareness, political will and public participation is essential for the successful implementation of the legal provisions and to have an integrated approach towards sustainable management of solid wastes in the country.

References

Central Pollution Control Board (CPCB) (2000), "A Support Manual for Municipal Solid Wastes (Management and Handling) Rules, 2000", CPCB, July, 2003 (Series: IMPACTS/11/2003-04).

Central Pollution Control Board (CPCB), 2006, "Assessment of Status of Municipal Solid Waste Management in Metro Cities and State Capitals", CUPS/65/2006-07.

Da Zhu, P. U. Asnani, Chris Zurbrügg, Sebastian Anapolsky and Shyamala Mani, (2008),"Improving Municipal Solid Waste Management in India-A Sourcebook for Policy Makers and Practitioners", World Bank

Hanrahan.D, Srivastava S and Ramakrishna A.S, (2006) "Improving Management of Municipal Solid Waste in India - Overview and

Challenges", Environment Unit South Asia Region, The world Bank.

http://envfor.nic.in/legis/hsm/mswmhr.html

Ministry of Environment and Forests, (2010),"Report of the Committee to Evolve Road Map on Management of Wastes in India", Ministry of Environment and Forests, New Delhi, March.

Chapter 3

Environmental Policies and Laws: A Legal Panorama

Gaurav Tiwari & Neha

The activist is not the man who says the river is dirty. The activist is the man who cleans up the river.

~Ross Perot

Introduction

Rapid and extensive industrial expansion, unplanned urbanization, migration of persons from rural to urban areas and improper and inadequate action of the authorities entrusted with the work of pollution control and environmental protection have largely contributed to unhealthy and degraded environment in India. This all, in turn, affected the quality of life of the large number of persons. Unplanned and alarming rate of urbanization has given rise to many environment related problems, such as problem of health and hygiene, sewage, disposal of solid waste, air, water and land pollution, slums, housing and basic amenities. Apart from air and water pollution, most gigantic problem which modern Indian cities are facing is the solid waste disposal problem. India is a *"Soft State"*. It has excellent environmental laws, but politics and/or corruption prevent their effective implementation.

Solid Waste

Any materials which are thrown away because of no value are called *waste such as* materials like garbage, sludge, biomedical waste etc. According to *Purden & Anderson (2008), 'refuse* and solid waste are about the same thing, *Garbage* is food waste, *Trash* and *Rubbish* are roughly equivalent terms; they contain little or no garbage. Trash frequently refers to grass and shrubbery clippings, papers, glass, cans and other household wastes. Rubbish is also likely to include demolition materials like brick, broken concrete, and discarded roofing and lumber.' It has further been clarified that solid waste includes discarded and abandoned appliances tank autos, mine and manufacture waste, agricultural, lumbering, hospital, research laboratories and industries. Some industrial wastes are toxic or hazardous.

In legal sense, **The Municipal Solid Wastes (Management and Handling) Rules, 2000** have defined *'municipal solid waste'* as 'commercial and residential wastes generated in municipal or notified areas in either solid or semi-solid form *excluding* industrial hazardous wastes but includes treated bio-medical wastes."

Constitutional Provisions

The liberal explanation of *"life"* in Article 21 has led to constructive progress of an environmental jurisprudence in India. On issue of correlation between ecology and Article 21, the opinion of the Court is that as the right to life is a Fundamental Right under Article 21, and given that the right to life connotes *"quality of life"*, a person has a right to enjoy pollution free water and air to benefit from the life completely.

The Apex Court has established the *doctrine of public trust* which lies on the principle that certain natural resources like air, sea, waters are resources for universal use and cannot be limited to personal ownership. These resources are the bequest of nature and the State, as a trustee thereof, is bound by obligation to preserve them. The State is the trustee, and general public the beneficiary, of such natural resources as sea, running waters, air, forest, ecologically fragile lands.[1] Environmental, ecological, air water pollution etc. add up to infringement of Article 21. Clean environment is consequently a vital truth of the right to healthy life as it is not feasible to exist with human dignity without a gentle and healthy environment. There is, for that reason, a legitimate imperative on the government not only to guarantee and sustain proper environment but also to acquire ample measures to encourage, guard and perk up both the man-made and the natural environment.

The Supreme Court has laid down that the *"Precautionary Principle"* and the *"Polluter Pays Principle"* are indispensable attributes of "sustainable development."[2] These notions are ingredient of Environmental Law of the country. " The 'polluter pays' principle as construed by Supreme Court connotes that the absolute liability for damage to the environment expands not merely to reimburse sufferers but also the cost of restoring the environmental degradation"[3] The precautionary principle signifies that the state government and the concerned statutory authorities must foresee, check and attack roots of environmental degradation.

Is Environment protection the duty of the State?

The respond to this question is Article 48A[4] of Indian Constitution which compels the state to undertake to guard and preserve the environment and to protect the forests and wildlife of the country. The Supreme Court has elucidated that at any time a quandary of ecology is produced before the Court, it is supposed to keep in mind Articles. 48A and 51A (g)[5] and cannot abscond the issue utterly to the government. To facilitate efficient steps taken for the rationale, wildlife and forests have now been positioned in the Concurrent List (See end note 24 for details) so that the Central Government may partake a significant task in this noteworthy locale.

In *M.C. Mehta v. Union of India*[6] the Court has held: "Article 39(e)[7], 47[8], 48A by themselves and jointly throw duties on the state to protect the health of the populace, improve public health and look after and improve the environment". With respect to Article 51 A (g)[9] inflicts a fundamental duty on each citizen and envisions that each citizen must execute their duties in an admirable approach and not lackadaisically.

Criminal Laws and Solid Waste Management

There are two major criminal laws dealing with solid waste management—(a) The Indian Penal Code (IPC), 1860; and (b) The Criminal Procedure Code,(Cr.PC) 1973.

The Indian Penal Code and Solid Waste Management

The Indian Penal Code of 1860 has dealt with solid waste management under Chapter XIV 'of offences affecting the public health, safety, convenience, decency and morals'. Since, solid waste gives rise to various types of diseases and is dangerous to public health; it has been treated as 'public nuisance' and has been made punishable. But there is no direct section in the Code which deals with the problem of solid waste.

Provisions under the Criminal Procedure Code, 1973

Section 133[10] of the Criminal Procedure Code, 1973 deals with 'removal of nuisance' and empowers the Sub-Divisional Magistrate or any executive Magistrate, on receiving report/ information, to make order to remove the public nuisance and desist from carrying any trade, business which is causing public nuisance. The Court has made use of Section 133 of the Code widely to deal with the problem of solid waste management. In the famous case of *Municipal Corporation, Ratlam vs. Shri Vardhichand* Justice Krishna Iyer declared that '...the guns of Section 133 go into action wherever there is public nuisance. The public power of the Magistrate under the Code is a public duty to the members of the public who are victims of the nuisance.' If the order is defied or ignored, Section 188[1111] 188:

Disobedience to order duly promulgated by public servant

Whoever, knowing that, by an order promulgated by a public servant lawfully empowered to promulgate such order, he is directed to abstain from a certain act, or to take certain order with certain property in his possession or under his management, disobeys such direction,

Shall, if such disobedience causes to tender to cause obstruction, annoyance or injury, or risk of obstruction, annoyance of injury, to any persons lawfully employed, be punished with simple imprisonment for a term which may extend to one month or with fine which may extend to two hundred rupees, or with both;

And if such disobedience causes or trends to cause danger to human life, health or safety, or causes or tends to cause a riot or affray, shall be punished with imprisonment of either description for a term which may extend to six months, or with fine which may extend to one thousand rupees, or with both., I.P.C. comes into penal play. It was further, observed that 'imperative tone of Section. 133, Cr.P.C. read with the punitive temper of Section. 188 I.P.C. makes the prohibitory act *a mandatory duty*.'

Public Interest Litigation: A Remedy

Prior to 1980s, only the aggrieved party could personally knock the doors of justice and seek remedy for grievance and any other person who was not personally affected could not do so as a proxy for the victim or the aggrieved party. But around 1980, the Indian legal system, particularly the field of environmental law, underwent a sea change in terms of discarding its moribund approach and instead, charting out new horizons of social justice. This period was characterized by not only administrative and legislative activism but also judicial activism.

In a modern welfare state, justice has to address social realities and meet the demands of time. Protection of the environment throws up a host of problems for a developing nation like ours. Administrative and legislative strategies of harmonization of environmental values with developmental values are a must and are to be formulated in the crucible of prevalent socio-economic conditions in the country. In determining the scope of the powers and functions of administrative agencies and in striking a balance between the environment and development, the courts have a crucial role to play. The judiciary's anxiety for combating environmental assaults has already been well elucidated. Its concern for the maintenance and preservation of forests, one of our depleting natural resources has also been highlighted.

Public Interest Litigation (PIL) has come to stay in India. According to Black's Law Dictionary[12] *"Public Interest Litigation means a legal action initiated in a court of law for the enforcement of public interest or general interest in which the public or class of the community have pecuniary interest or some interest by which their legal rights or liabilities are affected."*

Contrary to the past practices, today a person acting *bona fide* and having sufficient interest can move the courts for redressing public enquiry, enforcing public duty, protecting social and collective rights and interests and vindicating public interest. In course of time there has been a wave of environmental litigations.

At present most environmental actions in India are brought under Articles 32[13] and 226[14] of the Constitution. The writ procedure is preferred over the conventional suit because it is speedy, relatively inexpensive and offers direct access to the highest courts of the land. Nevertheless, class action suits also have their own advantages. The powers of the Supreme Court to issue directions under Article 32 and that of the high courts under Article 226 have attained greater significance in environmental litigation. Courts have made use of these powers to remedy past mala-fides and to check immediate and future assaults on the environment.

The formulation of certain principles to develop a better regime for protecting the environment is a remarkable achievement. In the Bhopal Gas case, the Supreme Court formulated the doctrine of *absolute liability* for harm caused by hazardous and inherently dangerous industries by interpreting the scope of the power under Article 32 to issue directions or orders which ever may be appropriate in appropriate proceedings. According to the Court, this power could be utilized for forging new remedies and fashioning new strategies.

These directions were given by courts for disciplining the developmental processes, keeping in view the demands of ecological security and integrity. In one of the earlier cases, *Rural Litigation & Entitlement Kendra vs. State of U.P*[15] , that posed an environment development dilemma, Supreme Court gave directions that were necessary to avert an ecological imbalance, such as constitution of expert committees to study and to suggest solutions, establishment of a monitoring committee to oversee afforestation programmes and stoppage of mining operations that had an adverse impact on the ecology.

In *L.K.Koolwal v. State of Rajasthan*[16], the Rajasthan High Court observed that a citizens duty to protect the environment under Article. 51-A(g) of the Constitution bestows upon the citizens the right to clean environment. The judiciary may go to the extent of asking the government to constitute national and state regulatory boards or environmental courts. In most cases, courts have issued directions to remind statutory authorities of their responsibility to protect the environment. Thus, directions were given to local bodies, especially municipal authorities, to remove garbage and waste and clean towns and cities.

In *Indian Council for Environ-legal Action vs. Union of India*[17], Supreme Court felt that such conditions in different parts of the country being better known to them, the high courts would be the appropriate forum to be moved for more effective implementation and monitoring of the anti-pollution law.

The liberal use of PIL against assaults on the environment does not mean that the courts, even if it is tainted with bias, ill will or intent to black mailing will entertain every allegation. This amounts to vexatious and frivolous litigation. When the primary purpose for filing a PIL is not public interest, courts will not interfere. In *Subhash Kumar vs. State of Bihar*[18], the Supreme Court upheld that affected persons or even a group of social workers or journalists, but not at the instance of a person or persons who had a bias or personal grudge or enmity could initiate PIL for environmental rights. The apex court in landmark judgment of *S.P.Gupta vs. Union of India*[19], elucidated in the following words: "*but we must hasten to make it clear that the individual who moves to court for judicial redress in cases of this kind must be acting bona fide with a view to vindicating the cause of justice and if he is acting for personal gain or private profit or out of political motivation or other oblique consideration, the court should not allow itself to be activised at the instance of such person and must reject his application at the threshold*".

The right to humane and healthy environment is seen indirectly approved in the MC Mehta group of cases (in mid 1980s), decided subsequently by the Supreme Court. The first MC Mehta case enlarged the scope of the right to live and said that the state had power to restrict hazardous industrial activities for the purpose of protecting the right of the people to live in a healthy environment. Although the second MC Mehta case modified some of the conditions, the third *MC Mehta v. Union of India*[20] case posed an important question concerning the amount of compensation payable to the victims affected by the leakage of oleum gas from the factory. The Court held that it could entertain a petition under Article 32 of the Constitution and lay down the principles on which the quantum of compensation could be computed and paid. This case is significant as it evolved a new jurisprudence of liability to the victims of pollution caused by an industry engaged in hazardous and inherently dangerous activities. The fourth *MC Mehta vs. Union of India*[21] case was regarding the tanning industries (located on the banks of Ganga) were alleged to be polluting the river. The Court issued directions to them to set up effluent plants within six months from the date of the order. It was specified that failure to do so would entail closure of business.

The four MC Mehta cases came before the Supreme Court under Article 32 of the constitution on the initiative of the public-spirited lawyer. He filed the petitions on the behalf of the people who were affected or likely to be affected by some action or inaction. The petitioner had no direct interest in the subject and had suffered no personal injury. Still standing to sue was not raised at the threshold question to be decided by the Court.

The Supreme Court has further expanded Right to life in recent years. In *Consumer Education and Research Centre vs. Union of India,* the Court said, Social security, just and humane conditions of work and leisure to workmen are as a part of his meaningful right to life?

The court held that this fundamental right to health and medical aid should continue even after retirement. Significantly, the Court said that in appropriate cases, appropriate directions could be issued to the state or private employer with a view to protecting the environment, preventing pollution in the workplace safeguarding the health of the workmen or preserving free and unpolluted water for safety and health of the people. Directions were issued to the Asbestos Industry, and the Union and State Authorities are meant to fill up the yawning gaps in the interpretation of the law.

The concept of compensation for environmental degradation has evolved at a snails pace over a period. It started with the strict liability principle followed by the absolute liability principle and then compensation under Article 32 and finally the " Polluter Pays Principle".

The Polluter Pays Principle means two things:

1. The polluter should pay for the administration of the pollution control system;
2. The polluter should pay for the consequences of the pollution.

This concept was further elaborated in the *Vellore Citizens' Welfare Forum vs. Union of India*[22], as follows:

The Polluter Pays Principle as interpreted by this court means that the absolute liability for harm to the environment extends not only to compensate the victims of pollution but also the cost of restoring the environmental degradation. Remediation of the damaged environment is part of the process of Sustainable Development and as such the polluter is liable to pay the cost to the individual sufferers as well as the cost for reversing the damaged ecology.

The society shall have to prosper, but not at the cost of the environment and in the similar vein, the environment shall have to be protected but not at the cost of development of the society. The need of the hour is to strike a balance between the two i.e., development on one side and pollution free environment on the other. A process by which development can be sustained for generations by improving the quality of human life while at the same time living in harmony with nature and maintaining the carrying capacity of life supporting eco-system. It focuses at integration of developmental and environmental imperatives. Thus, sustainable development is the only answer and administrative actions ought to proceed in accordance therewith and not abors the same.

Conclusions and Suggestions

The above discussion depicts that the problem of solid waste has not been considered as health & hygiene problem which ultimately affects the environment of the area. Though various laws were available but either they are not well couched to meet such type of exigencies, or were not efficiently and efficaciously implemented. Rather these laws were never made use of to deal with the problem of municipal solid waste. Because of inefficacious laws and inaction on the part of the implementing authorities, the Supreme Court was compelled to give serious thought to the problem and had to come out with detailed guidelines and the mechanism to overcome this menacing problem. It is clear that public at large is also not aware of inimical effects of this problem as it appeared in *Love Canal incident*[23] of the USA. It is getting less attention than required. Moreover, the landfill sites with toxic and hazardous wastes are like time bombs, they must be dealt with before they become 'foremost environmental problem' of this decade. In such cases one has to adopt 'cradle to grave' solutions from generation to final disposal of the wastes. This is particularly essential with the waste including toxic and hazardous wastes. To contain and control the problem, there must be '*awareness program*' to reduce solid waste, to reuse the things if possible and to recycle the wastes like paper waste. Thus, it deals attitudinal change amongst the people before we venture to strengthen the existing laws and implement them effectively and efficiently.

Reference

Purden and Anderson (2008) , "Environmental Law in India", Lexis Nexix Butterworths Wadhwa, Nagpur

Endnotes

1. M.C. Mehta vs. Kamal Nath, (1997) 1 SCC(Supreme Court Cases), 388.
2. M.P.Jain, Indian Constitutional Law, Edn. 2009, Lexis Nexix Butterworths Wadhwa, Nagpur p. 1139.
3. M.C. Mehta vs. Union of India, AIR 1997 SC 761.
4. Protection and improvement of environment and safeguarding of forests and wild life.
5. To protect and improve the natural environment including forests, lakes, rivers and wild life, and to have compassion for living creatures.
6. All India Reporter 2002 pg.1696.
7. Certain principles of policy to be followed by the State : *(e)* that the health and strength of workers, men and women, and the tender age of children are not abused and that citizens are not forced by economic necessity to enter avocations unsuited to their age or strength.
8. Duty of the State to raise the level of nutrition and the standard of living and to improve public health.

9. Fundamental duties *(g)* to protect and improve the natural environment including forests, lakes, rivers and wild life, and to have compassion for living creatures.

10. Examination of complainant. 133. —(1) When a Magistrate takes cognizance of an offence on complaint he shall at once examine the complainant upon oath.

(2) The substance of the examination shall be reduced to writing and shall be signed by the complainant and also by the Magistrate.

(3) This section shall not apply to —

(a) a complaint made by a police officer or by a public servant as defined in the Penal Code acting in his public capacity in respect of an offence punishable with imprisonment for a period not exceeding 6 months or with fine only; or

(b) a complaint by a public officer in respect of any offence against any law relating to local government or any by-laws or rules made thereunder for the time being in force, provided that the complaint is in writing and signed by the police officer or public servant.

11. 188: Disobedience to order duly promulgated by public servant Whoever, knowing that, by an order promulgated by a public servant lawfully empowered to promulgate such order, he is directed to abstain from a certain act, or to take certain order with certain property in his possession or under his management, disobeys such direction, Shall, if such disobedience causes to tender to cause obstruction, annoyance or injury, or risk of obstruction, annoyance of injury, to any persons lawfully employed, be punished with simple imprisonment for a term which may extend to one month or with fine which may extend to two hundred rupees, or with both; And if such disobedience causes or trends to cause danger to human life, health or safety, or causes or tends to cause a riot or affray, shall be punished with imprisonment of either description for a term which may extend to six months, or with fine which may extend to one thousand rupees, or with both.

12. Garner, Black's Law Dictionary, 8th Edition (2008).
13. Remedies for enforcement of fundamental rights where a person can directly file a suit in Supreme Court.
14. Every High Court shall have power, throughout the territories in relation to which it exercises jurisdiction, to issue to any person or authority, including in appropriate cases, any Government, within those territories directions, orders or writs, including writs in the

nature of *habeas corpus, mandamus*, prohibition, *quo warranto* and *certiorari*, or any of them, for the enforcement of any of the fundamental rights and for any other purpose.

15. AIR 1985 SC 652 (All India Reporter 1985 Supreme Court p.652).
16. AIR 1988 Raj.2(All India Reporter Rajasthan High Court p.2).
17. AIR 1996 SC 1446.
18. AIR 1991 SC 420.
19. AIR 1982 SC 149.
20. 1987 AIR 1086 1987 SCR (1) 819.
21. AIR 1988 SC 1115.
22. AIR 1996 SC 2715.
23. Neighbourhood in Niagara Falls, New York the site of the worst environmental disaster involving chemical wastes in U.S. history. Originally the site of an abandoned canal, it became a dumping ground for nearly 22,000 tons of chemical waste in the 1940s and '50s. The canal was later filled in, and housing was built on it. The leakage of toxic chemicals into these homes was detected in 1978, and residents were disupreme courtovered to have a high incidence of chromosome damage. After their evacuation, 1,300 former residents obtained a $20 million settlement from the dumping company and the city. In the early 1990s New York state ended its cleanup and declared parts of the area safe for residence.
24. Centre and State governments both have powers to enact laws. There are three lists in Indian Constitution namely, Union list: under which centre has the power to make laws. Secondly, State list: under which states have power to enact laws. Lastly, Concurrent list: under the subjects mentioned therein, both Centre & State can make laws.

Chapter 4

Vermicompositing Bioconversion Technology Through the use of Bio-Solids and Green Waste

Shagufta Jabin ,Amita Tayal and Ria Nanda

Introduction

There are large numbers of sludge produced continuously in most waste water treatment operation. The produced sludge needs to be deal with full care. Concerns about environment quality have lead to the introduction of alternative disposal methods such as the use of nutrient source for plants and soil conditioners[1]. The importance of biological processes in the management of sludge has been widely recognized within the broad range of bio-processes available. Composting and bio-composting are the most efficient techniques for converting sludge into useful products. The textile[2] and other industries wastes have a variable composition and normally contain high organic matters. During composting, the nutrients present in the wastes are converted by microbial action into soluble forms available to plants. Composting can also decrease toxicity of sludge[3].

Vermicomposting is a process where earthworms and micro-organisms convert organic material into nutrient rich humus called Vermicompost[4]. The earthworms fragment the organic waste substances, stimulate microbial activity greatly and increase the rate of mineralization rapidly converting the waste into humus like substances. The must condition in the Vermicomposting process is the maintenance of temperature below 35^0C because exposure of worms above this temperature kill them.

The integrated system from both thermophilic composting process and the Vermicomposting process would be necessary to provide a product, free of pathogens and with desirable characteristics at the faster rate than either of the individual processes. Furthermore, the combined technique accelerates the process and thereby reduces the time required.

Green waste[5] is bio-degradable waste that comprised of garden or park waste such as grass or flower cuttings and hedge trimming. Green waste is often collected in municipal kerbside collection scheme or through private waste management. The green waste fraction waste stream is important as biodegradable waste.

Processing details

The processing details of comprehensive technologies are as follows:

1. The sludge and other bio-solids[6] waste are collected from the different industries. The above wastes were taken into cylindrical shaped tank and are covered with fixed type of roof. It undergoes into anaerobic digestion. Anaerobic digestion[7, 8] is promoted by maintaining optimum temperature. In the digester, the complete mixing of the digested sludge in high rate digester is of homogenous nature. The processing details provides for short term compositing processing throughout the use of activated bacteria present in digested sludge[9].

2. To initiate proceeding in Vermicomposting channel, the channel is provided with a combination of green waste (20%), canning waste (30%) and dairy waste (50%). Initially, this channel will also be provided with earthworms suitably seeded. This initial processing channel should be two in numbers and the processing time will be 5 days and 4 days respectively.
3. Such initial processed material shall be transferred to channel 3 and 4 respectively for intermediate processing and contents would be properly churned manually but with least disturbance to earthworm.
4. In this intermediate processing channel, the earthworm will begin to come up and will flourish by suitable multiplication. This will help in the preparation of desired Vermicompost.
5. The intermediate products will be transferred to final vermiculture channel where earthworm will go down and compost will come up. Parts of final products will be withdrawn from the outlet.

The initial processing time will be 5 days and intermediate processing time will be 4 days whereas the total processing time for the final product shall be 10 days. Two channels have been provided for initial and intermediate processes so that two combinations will simultaneously work. The earthworms will be recycled to initial processing unit. This cyclic operation will continue and the operation will work without any break which has been shown in the flowchart.

Conclusion and Discussion

The word comprehensive indicates the following three points:

- This system can accommodate different types of wastes in different proportions; like green waste, canning waste and dairy waste.
- The sludge[10] helps in completing the process in much shorter span than that of the conventional systems[11].

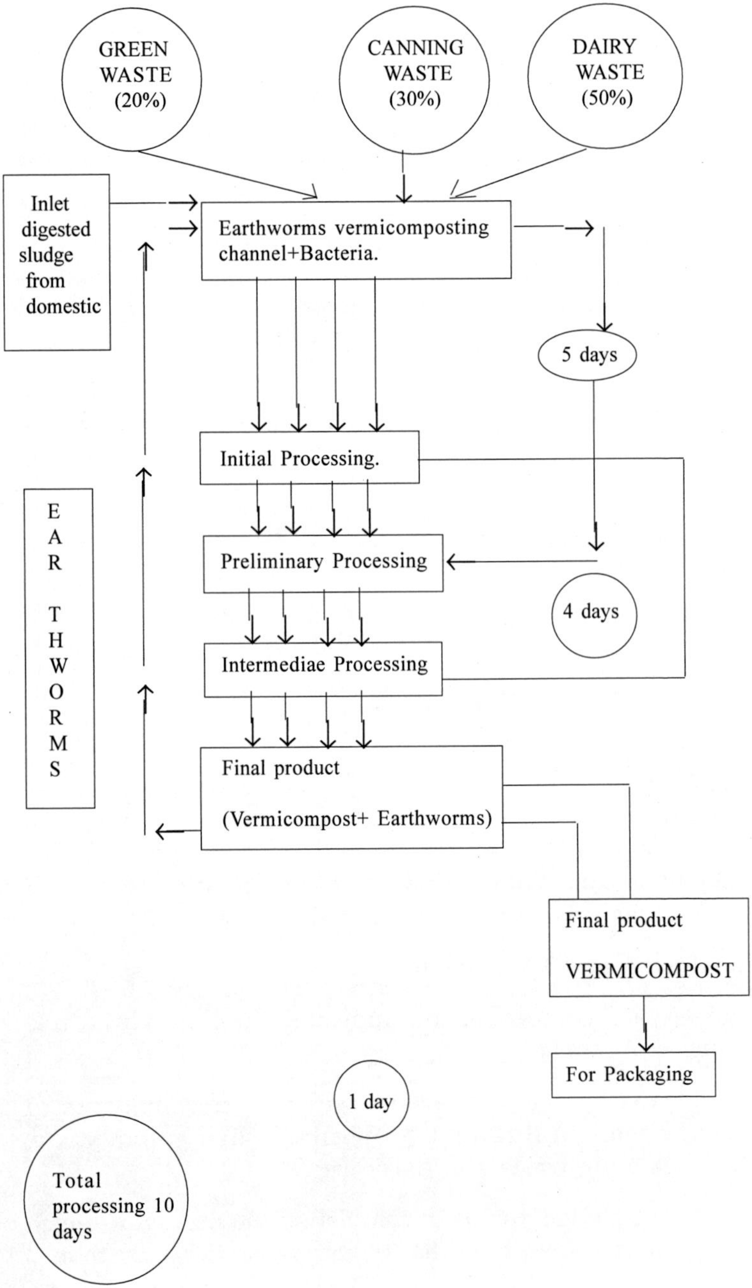
GREEN WASTE (20%)
CANNING WASTE (30%)
DAIRY WASTE (50%)
Inlet digested sludge from domestic
Earthworms vermicomposting channel+Bacteria.
5 days
Initial Processing.
Preliminary Processing
4 days
Intermediae Processing
E A R T H W O R M S
Final product
(Vermicompost+ Earthworms)
Final product
VERMICOMPOST
For Packaging
1 day
Total processing 10 days

- This technique works in different stages like initial, preliminary, intermediate and final stages; which again facilitates withdrawal at different intervals depending upon the quality of compost required or withdrawal in the final processing stage before sending in for final packaging.

The above work provides moderate thinking to increase the production while providing total flexibility at each stage. Assorted quantity of raw products and option of adding sludge for obtaining different qualities of finished products has never been tried earlier. Here the earthworm with mixed culture system has been tried for the first time. It was possible only because of introducing different processing stages so that bacteria and earthworm can play their role in vitality.

The results obtained indicate reduction in the amount of solid wastes with the comprehensive vermi-composting technology. It also indicates the reduction in the bio-degradable organic content and mineralization of sludge. The finding shows that comprehensive treatment of bio-solids with other wastes (viz. green waste, canning waste and dairy waste) is more efficient in stabilizing the sludge than the vermi composting process alone.

The flowchart presented above suggests that addition of green waste, dairy waste and canning waste gives richer quality organic manure which is more useful for the industrial purpose. It is possible to achieve only when the system is passed through multiple stages like initial, preliminary, intermediate and final processes. In addition the products obtained by the comprehensive system seem to be closer to the ideal standard of manure. The engineering of these biotechnology systems to efficiently and economically process a large volume of solid wastes has only recently been refined to the point of commercial viability.

References

1. Reddy, G.B., Cheng, C.N., Dunn, S.J., 1983, "Survival of Rhizbobium Japonicum in Soil-Sludge Environment", *Soil Biol Biochem*.15,343-345

2. Kaushik,P., Garg, V.K., (2003), "Vermicomposting of Mixed Solid Textile Mill Sludge and Cowdung with Epigeic Earthworm Eieinia Foetid",. *Biores. Technol.* 90,311-316.

3. Araujo, A.S.F., Monteiro, R.T.R., (2005), "Plant bioassays to Assess Toxicity of Textile Sludge Compost". *Sci.agri.* 62, 286-290.

4. U.S.Bhawalkar(1995),"Vermiculture Bioconversion of Organic Residue", *PhD thesis*, IIT, Bombay, India, 1995.

5. Frederickson, J.and Bult,K.R(1997),"Combining Vermin Culturewith Traditional Green Waste Composting System." *Soil biology and Chemistry*, 29: 725-730.

6. Balan, D.S.L., Monteiro, R.T.R., (2001), "Decolorization of textile Indigo dye by Fungi". J *Biotechnol.* 89, 141-145.

7. Ranade,D.R., Gadre,R.V.(1988), "Microbiological aspects of Anaerobic Digestion Laboratory Manual", *Maharashtra Association for the cultivation of Science research Institute*, Pune 1988.

8. Usepa(1994), "Sludge handling and Disposal Composting Federal Regulation", Washington, D.C.

9. Masciandra, G., Ceccati,B. and Gracia, C.(2000). "Insitu Vermicomposting of Biological Sludges and Impacts on Soil Quality". *Soil bio and Biochem* 32: 1015-1024.

10. Ayuso,M., Pascual, J.A., Garcia,C. and Hernomdez, T. (1996). "Evaluation of Urban Wastes for Agriculture Use". *Soil Science and Plant nutrition*, 42: 105-111.

11. Tomati. U., Grappelli,A. and Galli, E.(1987), "The Presence of Growth Regulators in Earthworm Worked Waste" in A M Bonvicini-Pagliai and Pomodeo on "Earthworm. *Mucchi Editor", Modena Italy*, 1987 ,423-425

Chapter 5

Technological Options for Municipal Solid Waste (MSW) Treatment and Disposal

Faisal Zia Siddiqui and Sourabh Manuja

Introduction

The Municipal Solid Waste (MSW) Management and Handling Rules, 2000 has specified the Muncipal Solid Waste (MSW) handling and management in four schedules. Schedule-I prescribes the implementation schedule for setting up of new MSW processing and disposal facilities and monitoring the performance of old MSW processing and disposal facilities. This also includes schedule for improvement of existing landfill sites and identification of new landfill sites for operation. Schedule-II prescribes the compliance criteria for management of different parameters like collection, segregation, storage, transportation, processing and disposal of MSW. With regard to treatment, the rules recommend adoption of a suitable technology or a combination thereof with the objectives of making use of wastes and to minimise burden on the landfill. For the bio-degradable fraction of waste the rules recommend treatment by composting, vermicomposting, anaerobic digestion or any other appropriate biological process so as to stabilise it. In other cases, options of incineration with or without energy recovery and pelletisation are also suggested. For the last element in the integrated chain, i.e., disposal, the rules recommend landfilling of only that type of waste which is neither recyclable nor biologically processable. The rules do not recommend landfilling of mixed waste as long as it is found to be suitable for any processing. Hence setting up a processing plant is considered to be a precondition for commissioning a sanitary landfill site. Schedule III prescribes the specifications for landfill sites covering: Site selection criteria, facilities to be provided at the site, specifications for land filling, provisions for pollution prevention, water and air quality monitoring requirements, plantations to be maintained at the landfill site, closure of landfill sites and post-closure plans. Schedule IV prescribes the standards for composting, treated leachates and incineration and operating and post closure standards.Although the rules on the one hand suggest this as an option while at the other place prohibiting disposal of bio-degradable waste in a landfill which is the basic element for landfill gas generation. Thus, the return on investment on landfill gas recovery may not be viable. In the implementation of these rules technology among others is a critical input. It is against this background that the present paper describes different MSW treatment and disposal technologies.

Characterization of MSW

Before identifying any MSW treatment technology, the characteristics of MSW should be established. The physical characteristics include size of the constituents, moisture content, bulk density, volatile solids, percentage of different components such as, paper, plastic, glass, metal, organic matter, inert contents, etc. The chemical characteristics include pH, percentage of nitrogen, potassium and phosphorus, fixed carbon, C/N ratio and calorific value. The identification of a suitable MSW treatment technology largely depends on the calorific value of the waste. The toxicity of the MSW is established by the concentrations of heavy metals, pesticides and insecticides in the waste. The factors which determine the potential for energy recovery from MSW includes the quantity, physico-chemical characteristics and the selection of treatment technology (Refer Table-1). Determination of carbon, hydrogen, oxygen, nitrogen and sulfur content (ultimate analysis) is done for both thermo-chemical and bio-chemical processes. In case of anaerobic digestion, the parameters C/N ratio (a measure of nutrient concentration available for bacterial growth) and toxicity requires consideration.

Table-1: Favourable range of important MSW parameters for energy recovery

Waste Treatment Method	Basic Principle	Important Waste Parameters	Desired Range
Thermo-chemical * Incineration	Decomposition of organic matter by action of heat	- Moisture content - Organic/Volatile	<45% >40%
* Pyrolysis * Gasification		- Fixed Carbon - Total Inerts - Calorific Value (NCV)	<15% >35% >1200 kcal/kg
Bio-chemical conversion *Anaerobic	Decomposition of organic matter microbial action	* Moisture content * Organic/Volatile matter	>50% > 40%
Biomethanation		*C/N ratio	25-30

Identification of Environmentally Sound Technologies (EST) for MSW Treatment

EST refers to cost effective and energy efficient technologies, which do not pollute the ecosystem's vital components such as air, land or water and consider the concept of 4 R's for MSW (Refer Table-2). EST can be categorised into:

- **Hard EST:** This includes equipment, machines and other infrastructure with their material accessories to handle solid waste products and monitor/measure the quality of air, water and soil.
- **Soft EST**: This supports and complements hard technologies and includes nature-based technologies and management tools. Nature-based technologies include processes and mechanisms nature uses within a specific ecosystem and its carrying capacity, while management tools include system, procedures, policies, regulatory frameworks, environmental performance standards and guidelines.

The EST can be selected based on the following generic criteria, the indicators of which may vary from region to region:

- **Affordability:** This means low investment, reasonability, maintenance-free and durability.
- **Validity:** This refers to effectiveness, ease in operation and its maintenance.
- **Sustainability:** This means low environmental impact, high energy efficiency and cultural acceptability

SWOT Analysis of MSW Treatment Technologies

SWOT Analysis may be conducted to assess various components of the MSW treatment and disposal options. The result of the analysis would provide insights towards developing a strategic plan for MSW management (Refer Table-3).

Table 3: SWOT Analysis of MSW Treatment Technologies

Issues	Compositing	Pelletisation	Biomethanation	Incineration	Gasification Pyrolysis	Incineration
Suitability to Indian waste	+ +	+	+	- -	- -	+ +
Experience in India	+ +	+	0	- -	- -	+
Resource recovery potential	+	+ +	+ +	+ +	+	-
Marketability of end product	-	+	+ +	+ +	+ +	0
Waste volume reduction	+	+ +	+ +	+ +	+ +	-
Capital investment cost	+	-	- -	- -	- -	+
O&M costs	+ +	+	- -	- -	- -	+
Ease in operation	+ +	+	- -	- -	- -	+ +
Suitability of treating mixed waste	+	- -	- -	-	- -	+ +
Land requirement	-	+ +	+ +	+ +	+ +	-.-
Environmental risks	-	-	+ +	-	-	- -
Scale of operation (larger the better)	-	+	+	+	+	+ +
Total (+)	12	12	12	9	8	11
Total (-)	3	4	8	12	13	6

Note: The "+" sign indicates positive impact; likewise, the "-" sign indicates negative impact.

Energy Recovery from MSW

The energy recovery from MSW is a viable option since the total quantity of MSW is reduced by 60 - 90%, depending upon the MSW composition and the technology adopted. The latent energy present in the organic fraction of the MSW can be recovered through the following methods:

- o **Thermo-chemical conversion** (Incineration/ Pyrolysis/Gasification) - entailing thermal de-composition of organic matter to produce either heat energy or fuel oil or gas; and are useful for MSW with low moisture content.

- o **Bio-chemical conversion** (Anaerobic Digestion or Bio-methanation) – entailing biological decomposition of organic matter by microbial action to produce methane gas, preferred for MSW having high percentage of bio-degradable matter and high moisture content.

Table 2: Comparative Evaluation of MSW Treatment Technologies

S.No	Item	Incineration	Pelletisation	Pyrolysis	Bio-Methanation	Landfill Gas generation
1.	Sorting and segregation	High	Very High	High	High	
2.	Energy recovery (in optimum conditions)	Approx 14 times	Approx 14 times	Approx 14 times	Approx 11 times	Approx 11 times
3.	Direct Energy Recovery	Yes	No	Yes	No	No
4.	Overall efficiency in case of a small set up	Low	Low	Moderate	High	Low
5.	Efficiency in case of high moisture	Very low	Very low	Low	Moderate	Moderate to High
6.	Land requirement	Low	Low	Moderate	Low to Moderate	High to very high
7.	Transportation costs	Moderate	High	High	High	Very high (depends on location of landfill)
8.	Potential for hazardous characteristics	Yes	No	Yes	No	No
9.	Toxicity of end product	High	NA	NA	NA	Moderate to High
10.	Leachate generation	None	None	None	High (in case of no protection layer)	High to Low
11.	Emission of Air Pollutants	High (not easy to control)	Moderate	Moderate	Low (easy to control)	Moderate
12.	Sustainability of source/ waste stream	Moderate	Low	Low	Low	High
13.	Initial Capital cost	High	Very High	Very High	Very High	High

Integrated MSW Treatment Technologies

Among the technologies mentioned above, the gasification and Pyrolysis technologies have not been considered as a appropriate option since they are still in R&D stage. Other technologies considered are: 1) Refuse Derived Fuel, RDF, 2) Anaerobic Digestion and Biogas Plant, 3) Sanitary Landfill and Landfill Gas Plant, 4) Incinerator Power Plant and 5) Compost. The Integrated MSW Management concept is used to manage the MSW appropriately according to its characteristics. Four options have been identified which have been described below:

1. RDF Plant and Sanitary Landfill

The Refuse Derived Fuel (RDF) is a renewable energy source. Other non-combustible and inert materials such as stone, ceramics, non-recyclables, and rejected waste from RDF production process shall be disposed off in a Sanitary Landfill. Recyclable materials such as metals and glasses shall be recycled. A precondition to this option is MSW segregation at source so that combustible and noncombustible materials shall not be blended, to ensure the quality of RDF.

Advantages of RDF

(a) High calorific value of pellets 3500-4000 kcal/kg against that of unprocessed garbage i.e.800-900 kcal/kg:

(b) Wherever necessary agro waste, industrial waste, plantation waste, forest waste etc., can be blended before pelletising:

(c) Pellets can be conveniently stored, transported and used as supplementary fuel for combustion processes, utility boilers and in industries such as cement kilns as a substitute to coal;

(d)Pellets have high thermal value and low ash content (2-10% as compared to 20-40% in coal).

Disadvantages of RDF

(a) The processing unit cannot be operated during rainy season, since MSW has high moisture content which increases the cost of drying.

(b) The energy consumption of the RDF facility is relatively high and the demand for maintenance of shredders, pelletiser and sorting equipment is substantial

(c) Alternatively, materials such as paper and plastics contained in the RDF fuel may have been recycled thus reducing the potential environmental and energy saving.

(d) The uncontrolled burning of the pellets derived from MSW may lead to harmful emissions. Further the inorganics are still to be disposed off and a landfill has to be operated.

2. Anaerobic Digestion, Composting and Sanitary Landfill

Since the composition of organic waste is high, the advantage is to convert organic waste to biogas by using anaerobic digestion technology. Biogas yield can be used as low cost fuel. As for organic wastes that are not suitable for the digester such as tree branches or grass, other rejects, and degraded sludge from the digester; they can be sent to compost plant to produce organic fertilizer / soil conditioner. Recyclable materials shall be recycled. Finally, the non-biodegradable and non-recyclable materials shall be disposed in the landfill. As a result, landfill in this option contains almost no organic wastes that can be further decomposed. Thus least leachate and odor would come out. Risk associated with groundwater contamination is thus reduced. This option segregated MSW at source, with appropriate standards to ensure the successful implementation of anaerobic digestion system.

Advantages of Anaerobic Digestion

(a) The option is suitable for MSW with high organic content;

(b) Compost/soil conditioner can be obtained as an end product of the process;

(c) The amount of waste for the landfill is considerably reduced;

(d) Landfill contains less organic waste compared to baseline thus reduces the opportunity of the groundwater contamination and green house gases emission. Further the closed system treatment reduces odor problems

Disadvantages of **Anaerobic Digestion**

(a) Combustible wastes cannot be used as fuel to produce energy ;

(b) The nature of organic waste may vary according to location and time of the year. In wet season horticulture waste may be higher than in the dry season. This may lead to a variation in the C/N ratio and can affect the rate of gas production;

(c) Inadequate mixing of MSW and sewage can affect the efficiency of the system;

(4) Not suitable for MSW containing less biodegradable material

Aerobic composting can be achieved by using micro-organisms such as bacteria, actinomycetes and fungi. In the initial stages, the bacteria and fungi activities cause the temperature in the compost plant to increase to 55-60°C, which is called the thermophilic stage. In the next stage, the bacterial and actinomycetes activity causes decomposition. This stabilized condition is known as mesophilic stage, when the composting material becomes dark brown due to humus synthesis. In addition to temperature, air supply, moisture content, particle size of MSW, acidity/alkalinity, and chemical characteristics are the other factors which influence the microbial activity in a windrow (Refer Table-4).

Table 4: Comparison of Composting Technologies

S.No	Parameter	Aerobic Composting	Anaerobic Composting	Vermicomposting
1.	Odorous Emissions	Yes	Yes	No
2.	Product Quality	Moderate	Good	Excellent
3.	Duration of Composting	Approx 14-21 days	Approx 180-240 days	180 days
4.	Production of Methane gas	No	Yes (under controlled environment)	No
5.	Nuisance of pests, flies, rodents etc	Yes	No	No
6.	Monitoring Mechanism	Not Constant	Constant	High consistency is required
7.	Storage capacity of end product	Low	Low	High
8.	Market demand	Moderate	Moderate	High
9.	Energy requirements	Yes	No	Yes
10.	Manpower requirements	Un-skilled	Skilled	Highly skilled
11.	Land requirement	Low	Moderate	High
12.	Sorting and segregation	Moderate	High	Very high
13.	Leachate generation	High	High	Low
14.	Soil & groundwater Contamination	High	Moderate	Low
15.	Initial Capital cost	Moderate	Moderate	High
16.	Operation & Maintenance Costs	Moderate	Moderate	High

3. Incineration Power Plant and Sanitary Landfill

In the case there is no system for MSW segregation, mass burning of all the wastes can be an option. The residue and ashes from incinerator are dumped in the landfill. The optimum calorific value of MSW should be approximately around 2000kcal/kg and also for the incinerator to operate sustainably. If the MSW has a calorific value of 500-600 kcal/kg, it is apparently not self-sustaining for an incinerator to burn the waste. Hence it is not possible to operate a waste to energy incineration system with MSW as the only input. The parameters controlling the incineration process includes Temperature, Residence Time, Turbulence Pressure, Air supply, Materials of construction, Auxiliary features, Waste moisture content, Heating value, High sulfur or halogens content. Further a proper incineration of MSW has certain features like Volume reduction, Stabilisation of waste, Recovery of energy from waste.

Advantages of Incineration

(a) The Volume and weight of the waste are reduced to a fraction of their original values (80 to 90%). Incineration requires a relatively small disposal area and the amount of waste for the landfill is thus reduced.

(b) The waste destruction is immediate; it does not require long-term residence time.

(c) If the ash residue and sludge become non- hazardous, incineration becomes a final disposal method as well as a treatment method for hazardous waste.

(d) The landfill contains less organic waste compared to baseline thus reduces the opportunity of the groundwater contamination.

Disadvantages of Incineration

(a) Metal fumes emission from the incineration process may be difficult to control / eliminate for inorganic wastes with heavy metal content (e.g. lead, cadmium, mercury, nickel arsenic. etc)

(b) Least suitable for high moisture content, low calorific value wastes and chlorinated wastes.

(c) Incineration represents a high capital cost. Auxiliary fuel is required to bring up an incinerator to operating temperature and in addition some other materials are required to maintain the combustion temperature.

(d) Besides particulates, SOx, NOx emissions, chlorinated compounds ranging from HCL to dioxins and heavy metals are a cause of concern, which require complex air pollution control equipments

4. Sanitary Landfill and Landfill Gas Production

This is the last option to be considered, but might only be needed in case improper segregation of MSW. All kinds of MSW are combined and it is difficult and costly to separate organic and non-organic components at the site. Therefore no further use can be drawn from those wastes.

Advantages of Landfilling

(a) Reduces GHG emissions;

(b) Landfill is flexible in receiving various kinds and amounts of waste;

(c) Landfill is the last treatment option of MSW management techniques;

(4) The system can produce landfill gas as a renewable energy.

Disadvantages of Landfilling

(a) Inefficient gas recovery process may yield only 30-40% of the total amount of gas actually generated, thereby increasing the GHG emissions from the remaining fraction.

(b) The landfill gas amount might be varied over time. Further the cost of pre-treatment to upgrade the quality of gas may be high.

(c) Spontaneous ignition/explosion may occur due to possible buildup of methane concentrations in the atmosphere.

(d) If the management is not up to sanitary standard, there is high opportunity of the groundwater contamination. system can produce landfill gas as a renewable energy

Conclusion

1. Waste reduction can be achieved by substituting reusable products for disposable products, patronizing renewable sources & biodegradable products and using products with greater durability.
2. Wherever it is economically viable and environmentally sound, the option for recycling of waste materials and products should be considered first. There should be a balance between waste minimization and waste recycling in order to minimize the environmental impact.
3. The concept of "waste to energy recovery" from MSW in India can be feasible if the waste is properly segregated at source or at the treatment and disposal facility.
4. Since the Indian MSW has a low calorific value, therefore "waste to energy recovery" plant has not been very successful in India.
5. In a landfill, the biodegradable components of MSW decompose to emit methane-a greenhouse gas 23 times more potent than carbon dioxide and may cause significant environmental problems. Therefore, the biodegradable component of MSW should be used for energy recovery from landfills.

References

Ministry of Urban Development (2000),"Manual on Municipal Solid Waste Management," Ministry of Urban Development, Government of India,.

MoEF (2000),"Municipal Solid Waste (Management and Handling Rules), 2000". Ministry of Envronment and Forests, GOI.

CPCB(2002) Municipal Solid Waste Processing Technologies, Reference Manual for local bodies, CPCB, CUPS/58/ 2002-2003.

MNEs "Energy Recovery from Urban Municipal and Industrial Solid Waste", MNEs.

Chapter 6

Role of Civic Bodies in Waste Management

S.A.M. Pasha

Waste Management impacts, in manifold ways, society, polity, environment, and economy of any country either in a positive or in a negative fashion. It traverses all areas of public and private spheres, and affects the daily activities of households and individuals. Having implications for environmental health, resource conservation and sustenance and the control of carbon emissions, waste management has, of late, begun to draw the attention of policy makers and civil society. Not to be left behind, to academia, it is a subject of burgeoning area of multi- and inter-disciplinary spheres of research. The disciplines of Economics and Geography, in particular, have taken up the subject as one of their major areas of theoretical and empirical research. Waste and its management draw into its ambit issues such as sustainable development, governance, consumption, environmental protection, and human rights with particular reference to right to health. The existence or the presence of an extensive body of regulations and statutory public authorities endowed with the principle of accountability to the public at large bringing within its vortex the notion of governance, are recognized as essential requirements for a proper and healthy management of waste in any polity, developing or developed. In the developed polities, interest in the subject (waste management) traversed a gradual line: a public health concern, concern for environmental protection, interest in resource conservation, and, of late, the carbon emissions implications. Guided by the dictum, "necessity is the mother of invention", knowledge and understanding of waste management have started progressing in consonance with problem identification.[1]

Waste Management Defined

Waste management is the collection, transport, processing, recycling or disposal, and monitoring of waste materials. The term usually relates to materials produced by human activity, and is generally undertaken to reduce their effect on health, environment or aesthetics. Positively constructed, waste management also involves the process of retrieving resources from it. Waste management can involve solid, liquid, gaseous or radioactive substances, with varying methods, and expert knowledge for each. Waste management practices, likewise, are distinct for developed and developing countries, for urban and rural areas within a country, and for residential and industrial generators of waste.

Means of Waste Disposal [2]

Landfill

Waste disposal in a landfill entails burying the waste. This is a commonly used practice in most countries. Landfills were often setup in abandoned or unutilized quarries, mining voids or borrow pits. A properly- designed and well- managed landfill can be a hygienic and relatively cheap method of disposing of waste materials. Older, ill- designed or badly-managed landfills can cause adverse environmental effects such as wind- blown litter, attraction of vermin, and generation of liquid leach ate.

Incineration

As a disposal method, Incineration involves combustion of waste material. Incineration and other high temperature waste treatment methods are also known as "thermal treatment". Incinerators convert waste materials into heat, gas, steam, and ash.

Individuals, on a small scale, and industries, on a large scale, can resort to Incineration. It is used to dispose of solid, liquid, and gaseous waste. It is recognized as a practical way of disposing of certain hazardous waste materials such as biological medical waste. One of the disadvantages of Incineration is that it may emit gaseous pollutants[3]. Among the countries where Incineration is common is land-scarce Japan, where such method does not require as much area as landfills.

Recycling

Recycling involves the widespread collection and reuse of everyday waste materials such as empty beverage containers. These are collected and sorted out into common types so that the raw materials from which the items are made can be processed into new products. Material for recycling may be collected separately from general waste using earmarked bins and collection vehicles, or sorted directly from mixed waste streams. The most common consumer products recycled include aluminum beverage cans, steel food and aerosol cans, HDPE and PET bottles, glass bottles and jars, paperboard cartons, news papers, magazines, and cardboard.[4]

Sustainability

Waste management is a key player in maintaining a business's ISO 14001 accreditations. Companies are encouraged to improve their environmental efficiencies annually. One way to do this is by improving a company's waste management with a new recycling service.

Biological Reprocessing

Waste materials that are organic/ biodegradable in nature, such as plant material, food scraps, and paper products, can be recycled using biological composting and digestion processes to decompost the organic matter. The resulting material is then recycled as mulch or compost for agricultural purposes. Besides, waste gas from the process (such as methane) can be captured and used for generating electricity. The purpose of biological processing in waste management is to control and accelerate the natural process of decomposting of organic matter.

Energy Recovery

The energy content of waste products can be harnessed directly by using them as a direct combustion fuel, or indirectly by processing them into another type of fuel. Recycling through thermal treatment ranges from using waste as a fuel source for cooking or heating, to fuel for boilers to generate steam and electricity in a turbine.

Avoidance and Reduction Methods

This method is akin to an adage, "prevention is better than cure". It involves the avoidance or prevention of waste from being created. Methods of avoidance encompass reuse of second- hand products, repairing broken items instead of buying new, designing products to be refillable or reusable such as cotton instead of plastic shopping bags, encouraging consumers to avoid using disposable products (disposable cutlery), removing any food/ liquid remains from cans, packaging.

Waste Handling and Transport

Waste collection methods vary widely across countries and regions. Domestic waste collection services are often provided by local government entities, or by private industry. In the developing world, the waste collection system is in doldrums.

Education and Awareness

Education and awareness in waste management is increasingly significant from a global perspective of resource management. Local, regional, and global air pollution; accumulation and distribution of toxic waste; destruction and depletion of forests, oil, and water; depletion of the ozone layer and emission of green house gases threaten the survival of mankind and thousands of other living organisms, the integrity of the earth and its biodiversity, the security of nations, and the heritage of posterity.

The coalescence of the aforecited means of waste disposal amount to what is described as Integrated Waste Management (IWM), a strategy increasingly adopted by the developed world to tackle the menacing affliction of waste. The IWM consists of a hierarchical and coordinated set of actions that reduces pollution, seeks to maximize recovery of reusable and recyclable materials and protects human health and environment.[5]

The increasing urbanization and fast- changing life styles visible in the urban areas in India, constitute, *inter alia*, the causal factors towards increasing volumes of waste. Conservative estimates tell us that Indian cities, at present, generate eight times more solid waste than what they used to at the time of independence. Alongside, waste generated per capita by an individual person is also increasing. Quantities of plastic material are seventy times higher than in the 1960s. Throughout the country, more than 1/3rd of all waste generated remains uncleared on the streets and what is collected is dumped in unsanitary dumpsites polluting ground water and degrading the environment[6].

The Role Of The Civic Bodies

The performance of the civic bodies in solid waste management in India is, to say the least, unsatisfactory. The improper management of solid waste cause air, land and water pollution, and poses risks to human health and the environment. Statutory bodies such as the Central Pollution Control Board need to come forward and play a pro-active role in ameliorating the existing state of affairs regarding waste management. Besides, the revamping of the transport sector and the state bureaucracy at all levels is the urgent need of the hour. A fast developing economy like India can now have a recourse to low-cost, low- tech, labour-intensive and decentralized waste management options available in the country.

For municipal authorities in India, waste disposal poses a challenging task testing their nerves and *raison dêtre,* in a very real sense. Waste disposal is an engaging assignment calling upon the attention of policy makers, regulators and civil society composed of concerned citizenry. Disseminators of knowledge and their potential role towards improving the situation pertaining to waste management can not be gainsaid.

Some of the recent empirical studies undertaken by conscious and concerned individual and civil society entities do unmistakably establish the yawning gap between the promises of the state bodies and their actual performance germane to the task of waste management. Writing on the administration of solid waste management in Delhi, Kaveri Gill observes:[7]

> For administrative purposes, the M[unicipal] C[orporation] of D[elhi] area has been divided into twelve zones, which are further divided into municipal wards, which are further broken down into beats. The institution [MCD] employs about 3700 safai karmacharis or sweepers as permanent employees of the government. It devotes approximately 15-20% of its operational budget to Solid Waste Management[SWM] and still *fails to serve half of the urban population, overlapping mostly with poor residents of slums*[emphasis supplied]. [8]

Continuing further, Kaveri Gill brings to the fore state's another sordid tale as regards its 'green' policy towards plastics:

> On the one hand the government fully condones the imports of plastic waste and on the other, they wish to ban the recycling of waste that is produced in our country [India]. Either we are producing too much waste, in which case the primary plastic producers and industrial users of plastic such as Pepsi, Coca Cola, and Bisleri should be held responsible, or a ban should be imposed on the imports of plastic waste from the developed countries .[9]

A mindset marked by duplicity or hypocrisy of the state is clearly visible in the aforesaid. Other studies on solid waste management attribute the crisis to the nature of developmental model adopted by the state. This is a viewpoint held by some of the non- state entities working in the field of waste management and are also engaged in the task of creating a conscious citizens body which in turn exert public pressure on the state. Toxics Links, a civil society organization, believes that the model of development that the world and India has followed has led to a situation where we are surrounded by numerous hazardous substances in our daily lives, effecting the health of people, of the living creatures, and of the planet as a whole. This is particularly borne out by e- waste emanating from the discarded components of computers and TVs and other electronic goods. [10]

By Way of Conclusion

Society-polity symbiosis is supposed to inform public affairs in democracies. Participatory democracy is being regarded as a panacea for all the ills besetting societies living under democratic regimes. Notwith standing this panacean conspectus among epistemic and policy communities, the reality is squarely different from these suppositions.

Waste management in developing countries in general and India in particular constitutes one of the crises of governance. Biodegradable and non- biodegradable waste management have to be meted out a distinct treatment in view of their different toxicities. We often hear in the Indian media that the electronic and nuclear wastes emanating from the advanced West are sought to be dumped in Indian soil for some pecuniary considerations. Such hazardous waste is capable of polluting this soil, besides posing a potential threat to human survival.

In the Indian polity, local self- governments in the current regime of panchayati raj institutions are entrusted with the task of ameliorating the conditions germane to waste management. The performance record of these entities remains much to be desired. Access to clean environment, ecology, air, and water are inextricably related to right to life. Adoption of viable waste management techonologies will go a long way towards improvement. Civic bodies can not be absolved of their obligations towards their citizens at the grass roots level. Empirical evidence unmistakably tells us that the involvement of women in civic administration in the developing world has brought to the fore positive outcomes. Their increasing involvement will further result in creative and desirable results. It is necessary to instill the sprit of stakeholdership among population at large. In an age of partnerships of various hues, citizen-civic entity partnership should form an essential component of any waste management governance. In the fast changing world of science and techonology, what we call 'waste' is no longer waste, as the latter is converted into energy in the West. It is not too difficult for a country like India to emulate the latter for it is well endowed in terms of scientific and technological manpower.

Endnotes

1. Pauline Deutz and Lynne E. Frostick, "Reconciling Policy, Practice, and Theorisations of Waste Management", The Geographical Journal, Oxford, UK, Vol. 175, No. 4, December 2009, pp. 247-250.
2. Wikipedia, the free encyclopedia, http.//en.wikipedia.org/wiki/Waste_management.
3. For opposition to Incineration project titled Clean Development Mechanism (CDM) of the United Nations, see a popular essay of John Vidal, "Waste- Pickers Oppose UN Plan", The Hindu, New Delhi, 6 August 2010, p.9.
4. For a legal regime governing packaging, see K. Vidyullatha Reddy, "Green Consumerism and Packaging Waste Management: The Indian Legal Scenario", Nalsar Law Review, Hyderabad, Vol. 4, No. 1, 2008-2009, pp. 160-168.
5. Mitrajit Chatterjee, " Integrated Waste Management- The Need of the Hour (With Special Reference To West Bengal)", Asian Studies, Kolkata, Vol. XXVII, No. 2, July- December 2009, p. 11.
6. Ibid., p. 9.

7. See, for details, Kaveri Gill, Of Povery and Plastic: Scavenging and Scrap Trading Entrepreneurs in India's Urban Informal Economy, New Delhi, Oxford University Press ,2010, p. 49. And also, Toxics Link, Our Toxic World: A Guide to Hazardous Substances In Our Everyday Lives, New Delhi, Sage Publications, 2010, pp. 180.

8. Ibid., p.49.

9. Ibid., p. 205.

10. A startling and a graphic account of the hazardous nature of e-waste on the health of the people who are engaged in the scrap business as appeared in a reputed, popular, New Delhi- based environmental magazine, Down to Earth. See its issue of May 16-31, 2010, Vol. 19, No. 1, pp. 22-32. Also see, P. Sree Sudha, "Tackling Electronic Waste: Need of the Hour!", NALSAR Law Review, Hyderabad, Vol 4, No.1, 2008-09, pp. 17-26.

Chapter 7

Capacity Building for Waste Management-- Toolkits for Preparation of Master Plan for Solid Waste Management

Ashfaque Alam

Background

Urbanization is now becoming a global phenomenon, but its ramifications are more pronounced in a country like India. Natural growth of population, reclassification of habitation and migration trends are important in urban population. The urban population in India was 285 million in 2001, which accounts for 27.80 per cent of the total population, it will be 41 per cent in 2021. Global experience shows that when a country's urban population reaches almost 25 per cent of the overall population (as in the case of India), the pace of urbanization accelerates. Indian urban population is increasing between 3% to 3.5% per cent per annum and yearly increase of waste generation is around 5 per cent. The per capita solid waste generated daily ranges from about 100 gram. in small towns to 500 gram. in large towns.

The scenario of economics of waste generation is quite interesting. The per capita income of developed countries is many folds higher than the developing and underdeveloped economies, thus increasing their purchasing power. Consequently, the waste generating ability and tendency in higher income groups is higher. On an average 25 per cent of the world population generates 80 per cent of the world's goods and services and consumes 85 per cent of its wealth and produces 80-90 per cent of solid waste. On the contrary the waste generating trends in the poverty driven countries is very less and account only about 10-20 per cent of the total solid waste.

Present Status of Waste Management

Presently the systems are assuming larger importance due to population explosion in municipal areas, legal intervention, emergence of newer technologies and rising public awareness towards cleanliness. The present status of municipal solid waste management is given below:

- Storage of waste at the source is lacking
- Domestic waste thrown on streets
- Trade waste on roads/streets
- Construction debris left unattended
- Bio-medical waste disposed in municipal waste stream
- Industrial waste disposed in open areas
- Segregation of recyclable waste at source not done
- Primary collection of waste not done at place of generation
- Design and location of municipal waste storage depots inappropriate, resulting in littering of garbage
- Street sweeping not done everyday
- Waste transportation done in open vehicles
- Waste processing partially practiced in limited number of ULBs only

- Final disposal done through crude dumping
- Rag pickers collect recyclables from municipal bins/ dumpsites and litter the waste causing insanitary conditions

Growing Problems due to Solid Waste

Solid wastes generated by domestic, commercial and industrial activities are often indiscriminately disposed. Unscientific management of such wastes leads to serious environmental problems. Insanitary method being adopted for disposal of solid wastes is a serious health concern. Particularly during rainy season, run-off and high humid conditions increase the health hazards. The land-fill sites, which are not well maintained are prone to ground water contamination due to leachate percolation. Open dumping garbage serves as breeding ground for disease vector such as flies, mosquitoes, cockroaches, rats and other pests.

Reasons for Improper Management of Waste

There are several reasons for improper management of solid waste. These are due to lack of:

- planning for waste management while planning townships;
- proper institutional set up for waste management, planning and designing in urban local bodies;
- technically trained manpower;
- community involvement;
- expertise and exposure to city waste management using modern techniques / best practices;
- awareness creation mechanism;
- management information systems;
- funds with ULBs; and
- Indifferent attitude of ULBs to levy user charges and sustainability

Need for Master Plan

Master Plan is a comprehensive document keeping in view of the long term requirement of the town. It contains: i) existing infrastructure available; ii) gaps and issues; iii) project service demands and targets to be achieved; iv) plan system; and v) various interventions and subprojects

To conceive, design, develop and implement the plan for improvement of Solid Waste Management system in a city on a long term basis in an organized manner, a document is required. This document will help in assessing the long term requirements both in physical and financial terms and shall act as a vision document for all future activities related to SWM sector. The Master Plan may be prepared with the objective to design a suitable integrated solid waste management system (ISWM), to assess the resource requirement for next 20 years and suggest suitable measures for improvement of solid waste management system. And to ensure compliance of Hazardous Wastes (Management and Handling) Amendment Rules, 2000 there is a need for Master Plan for SWM. Solid waste management is one of the eligible projects under Jawaharlal Nehru National Urban Renewal Mission (JNNURM), 2005.

Preparation Stages of Master Plan for Solid Waste Management

The following Stages will help to prepare the Master Plan for Solid Waste Management

(1) Define the project area

(2) Review of existing situation

(3) Service delivery norms

(4) Identification of deficiency areas

(5) Estimation of demand

(6) Preparation of plan

(7) Cost estimation

(8) O & M plan

(9) Institutional reorganization and capacity building

The above stages are explained below in details:

Stage 1: Define the Project Area

The project area may be defined in the context of geographical boundary of the town/city; mapping of the project area with ward boundaries and road network. The project area may be for define horizon year in accordance with the available Master Plan/Development Plan/Regional Plan. The project area may also define the major components of Solid Waste Management such as storage, street sweeping, collection, transportation, processing and disposal.

Stage 2: Review of Existing Situation

The existing situation of the project area may be reviewed in the context of:

i. General profile of the city/town with respect to its population, area, population density, important activities and communication networking;

ii. Existing service coverage by area and activity;

iii. Land use in the context of exiting development pattern, Master Plan proposals, future development orientation and any special activities like tourism, religious, etc.;

iv. Existing solid waste generation category and characterization of waste such as its density, moisture, caloric value, biodegradable, non-biodegradable etc., and total quantity generation per day;

v. Present system in operation by types of operation activities, infrastructure available, technology used and institutional set-up;

vi. Financial aspects will cover budget details for Municipal Solid Waste Management (MSWM) operation for previous years (last 3 years) and user charges details including rates, collection, efficiency etc.;

vii. Involvement of private sector in the context of area involved and the operation methods.

Stage 3: Service Delivery Norms

The service delivery norms may be monitored in the context of:

(i) Agencies governing SWM Norms;

(ii) Legal aspects (rules) published by Ministry of Environment and Forest (MoEF), Government of India (GoI) and

(iii) Norms/Guidelines for different activities such as sweeping, community bins, transport system, transfer station, processing plant and SLF (sanitary land fill).

The Norms/Guidelines for different activities such as sweeping, community bins, transport system, transfer station, processing plant and SLF are described below:

General Norms

MSWM on all seven days a week including street sweeping and shallow drain cleaning; 100% covered storage and transportation; processing of entire organic waste and disposal of only processed residual and non-organic waste

Specific Norms

Sweeping Norms - 250-350 MT for high density area having more than 50000 population/sq.km; 400-600 MT for medium density area having population between 10,000 and 50,000 sq. km; 650-750 MT for less than 10000/sq. km.

Community Bin Norms - Community bins as per utility and waste quantity; provision for storage of segregated waste; 100% more in numbers; distance of containers from work place < 250 m; inter bin distance < 500m; no bin in congested areas; litterbins in important areas of tourist importance

Transport System - Appropriate equipment of optimum capacity for segregated waste; open and slow moving vehicle to be replaced; 1 pushcart for each sweeper; 1 bin of 4.0 cum for 2000 persons; 1 carrier for 15 bins; 1 mini waste collector for 7 MT; 1 compactor for 60 MT etc; 1.2 cum space for every MT of waste etc. add 17% more for workers and 25% more for equipment.

Transfer station norms - If landfill > 10 km- Transfer Station (TS); for large quantity transportation- more suitable; large size containers of 15-20 cum capacity, work in two shifts; ramp with level difference, static compaction units; weigh bridge and other infrastructure as per requirement; better to take services of expert for designing and operation, management

Processing plant - Should be suitable - physical and chemical characteristics, relative assessment of disposal options; require minimum space, least initial capital; cause least adverse impact on environment, least residual generation, be sustainable-recover maximum net energy/ useful material, requiring least Operation and Maintenance(O& M); should be suitable for 20 years;2-3 acre for 50-75 MT; 2 Ha for 500 MT

Infrastructure include- internal road, drain, power, water supply, plantation, drain, weigh bridge, office, toilet blocks, parking for vehicles, shelter room for workers, leachate pond, compound wall and gates, security guard's cabin, machine shed, , compost pad, covered monsoon shed, curing shed, packaging and storage sheds etc (If site is part of SLF site, there could be common facilities and equipment)

Plant and machinery include- pre processing and processing machines etc., elevators, vibro screens, gravity separators, packaging sprouts, drives, other accessories and weigh bridge.

SLF site - Site should be sufficient for minimum 20 years: buffer zone of 500 m of no-development and proper land use; NOC from SPCB,ground water department, airport authority; environmental, social and geotechnical issues are to be addressed;

No landfill should be allowed within 200 m of any lake or pond; 100 m of river or stream; 100 year flood plain; 200 m of right of way; 300 m of public park; 500 m water supply well. Landfill site should atleast be 2 m above ground water surface and be away from airfields at least by 20 Km.

Should not be part of 10-year ground water recharge area; should be free from underground mines; seismic risks, away from habitation, forest, monument, water body, national park, wet land, any important place of historical/cultural or religious nature.

If site for processing plants are available near the city, then SLF sites could be away otherwise, processing plant and SLF site should be setup near to each other which would save transportation, operation and maintenance and supervision cost.

Stage 4: Identification of Deficiency Areas

The deficiency areas will be identified in the project area by evaluating existing system against the recommended norms with respect to the identified SWM components. Deficiencies may be list out by taking the information of manual and multiple handling, work distribution, improper and insufficient equipment, poor institutional setup, processing system and about SLF site. Due to deficiencies there is impact on town with respect to health risks (diseases), environmental risks (land, air, water pollution) and aesthetic risks (odor, visual, litter etc.) (Figure 1).

Figure 1: Risks Associated with Solid Waste Management

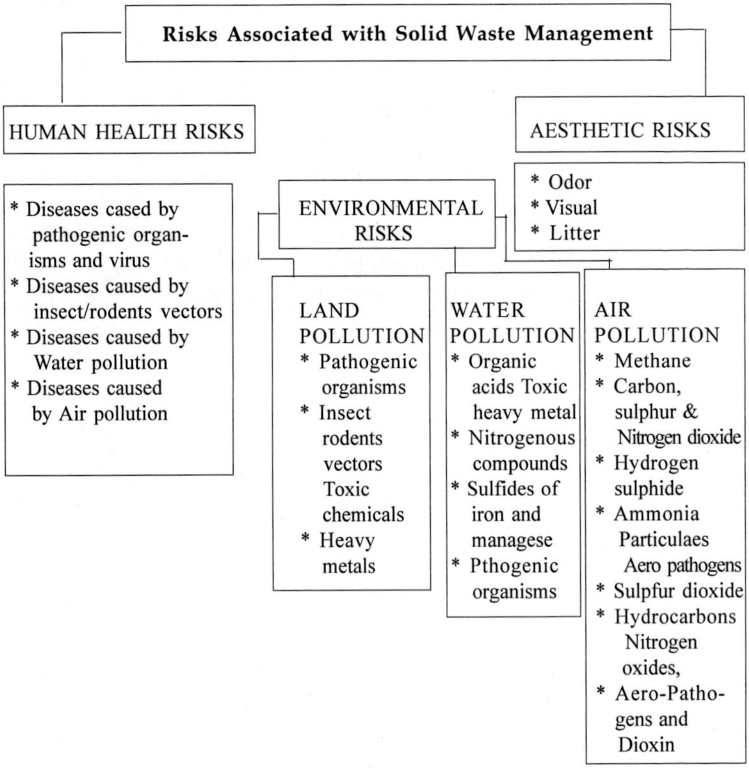

(Source: Policy Guidelines for Control of Environmental Pollution in Urban Areas of Developing Countries)

Stage 5: Estimation of Demand

The demand will be estimated in accordance with

(i) population projection;

(ii) estimate of future land use activities that generate solid waste (like multistory buildings, commercial activities, hospitals, industries, IT industries etc);

(iii) per capita waste generation;

(iv) estimation of waste quantity in each year;

(v) estimation of infrastructure requirements for different MSWM components after adjusting suitable technical options for primary collection, street sweeping, storage and

(vi) secondary collection system, transport to transfer station, transfer station construction and management, processing the waste (compost plant) and disposal at SLF site.

Work Norms for Workers for Primary Collection/Street Sweeping - As per Government order; 28 workers per 10,000 population has been prescribed. Normally, one sweeper should clean 1000 m of road length along with drains of shallow depth (up to 0.60 m) but in absence of road length information, government norms are adopted for preparation of the report. One push cart worker should be associated with two sweepers, thus number of pushcart holder will be half the number of workers.

Norms for Primary Collection Equipment - Every out of two workers should be allotted one 4/6 containerized pushcart. Number of pushcarts should be worked out by adding 25% spare equipment. Containerized mechanized four wheeler mini garbage collector is proposed for primary/door to door collection. These mini collectors should have separate compartments for biodegradable and non biodegradable waste with hydraulic unloading system and compatible with compactors as well as dumper placer containers thus avoiding the need for multiple handling. One fixed compartment is also provided for collection of recyclables. These vehicles should be used for door to door collection/primary collection and should ply upto Transfer station and unload the waste directly into compactors. One mini garbage collector should be able to collect about 7 Million Tonnes (MT)waste per day from 13000 persons.

Norms for Waste Storage Bins and Carrier Vehicle - There would be no need to provide community bins, if the primary collection service could be provided to 100% population. In that situation, only littered bins would have been kept in busy areas and the city would have become "bin free city". Community bins would also be required for collection of road swept material, littered material, bulk waste and waste from new areas where primary collection system would not be available. Normally, one bin of 4.5 cum size should be sufficient for a population of 4,000 persons but number of bins has been worked out @ one bin per 2,000 persons. One dumper carrier vehicle should make 8 trips in one shift or 15 trips in two shifts. One Compactor of 14 cum size should be sufficient for 5 mini garbage collectors or 35 MT per day.

Norms for Processing Plant and SLF site

Processing Plant:

Selection of suitable processing technology should be based on physical and chemical composition of waste, availability of land, capital cost, operation and maintenance cost, financial & economic viability including marketability of final product and sustainability aspects.

Site for processing plant should be sufficient for 20 years and be designed for entire biodegradable waste quantity from the entire city. Ideally it should be near to SLF site, so that residual could be disposed off without requiring any additional transportation cost.

Sanitary Land Fill Site

Land area should be large enough to last for minimum 20 years. If sufficient land is not available at one particular location, two-three locations in different sides of the city should be identified and developed which will save traveling cost and time. The land fill site should be away from habitation clusters, forest areas, water bodies, National Parks, Wet lands, Monuments and places of important cultural, historically or religious interest. A site accessible within 30 minutes travel time is suitable but for distance greater than 30 minutes travel time, large capacity collection vehicles or provision of transfer station should be made.

Situation of Landfill Sites - The land fill sites are also a crisis of availability of land in urban areas as due to rapid urbanization. According to the Central Pollution Control Board, the space required for land filling has increased by 20,000% in the last 50 years i.e. from 1947 to 1997. Around 90 per cent of the generated wastes are land-filled requiring around 1200 hectares of land every year with an average depth of 3 m. Due to rapid urbanization, prevailing land use regulation and completing demands for available land, it is desirable that adequate land be earmarked at the planning stage itself for solid waste disposal. The larger quantities of solid waste and higher degree of urbanization will necessitate better management involving a higher level of expenditure on manpower and equipment.

Stage 6: Preparation of Plan

This stage of Master Plan may consist the inputs of the infrastructure requirements to meet the target year demand under different components and specifications for the infrastructure requirements, equipments, its life cycle period etc.

Stage 7: Cost Estimation

Broad cost estimate for each phase comprising of all SWM components/elements, capital replacement plan as equipment will require replacement after end of economic life, primary collection equipment shall be replaced after 5 years, transport after 10 years and heavy machines like bulldozers/landfill compactors after 20 years. The cost will be estimated with respect to the above in addition of contingencies and allowance to total cost estimation.

Solid waste management involves capital investments and recurring expenditure. The possibility of revenue returns in the field of solid waste sector is not very encouraging in existing scenario, therefore, while taking decision about setting up of a processing plant, a realistic approach should be followed, some concessions and user recovery charges should seriously be implemented. Few of the cost related considerations to be taken, have been elaborated below:

- The annual expenditure includes payment of salary to staff, and operation and maintenance of equipment, machinery, vehicles, operation and maintenance of transfer station, processing plant and sanitary land fill site.
- A capital investment is required for the purchase of equipment, machinery and vehicles, setting up of a processing plant, setting up of a transfer station, acquisition of appropriate site for development as sanitary landfills and construction of new sanitary landfills.
- Additional haulage cost will be incurred when solid waste transportation network is changed. This would be different for various alternatives.

Addition of contingencies and allowance to total cost estimation may be based on - a) 5 per cent annual increment as price escalation on current rate, b) 3 per cent as physical contingency, c) 3 per cent for project preparation, d) 1 per cent each for social, environmental, IEC (Information, Education and Communication) and capacity building, and e) special provision for utility shifting and any other visible expenditure on project specific basis.

Stage 8: Operation and Maintenance (O & M) Plan

Operation and Maintenance plan will be in accordance with the identification of component specific O & M requirements; quantity and estimation cost requirements on annual basis; integration and utilization plan for existing facilities and municipal staff and possibility of private sector participation in O & M activities. The component specific of O & M identification may be based on primary analysis for each component-like for sweeping, collection and transportation equipment, transfer station, processing plant and SLF (Sanitary Landfill) site.

Stage 9: Institutional Reorganization and Capacity Building

In order to make city's solid waste management successful, involvement & efforts of line agencies becomes very important. Institutional strengthening should be done by adequately decentralizing the administration, delegating adequate powers at the decentralized level, inducting professionals into the administration and providing adequate training to the existing staff.

It is also necessary to fix work norms for the work force as well as for supervisory staff to maintain the manpower productivity and optimum output expected from the vehicles and machinery utilized. It is therefore, necessary that local body adopt adequate measures for institutional strengthening as mentioned above.

Institutional changes required with respect to institutional reorganization and capacity building.

A. Proposed Organizational Set-up

Exclusive SWM In charge from engineering background - Although, there is exclusive health officer to look after and coordinate the SWM activities in an exclusive manner but implementation is in traditional manner and not according to MSW (M&H) Rules, 2000. Normally, administrative officers and health officers are not experts in bidding procedures which compels them to continue with traditional style of working but for implementing good practices in SWM preparation of bid documents, tendering, formulation of specifications etc are required, which are related to engineering section but due to lack of coordination, lack of authority and personal conflict among officers most of the tender related activities have not been initiated.

A senior officer of Executive Engineer (X.EN) rank should be made the in charge of central SWM department. He should be responsible for entire implementation of SWM according to MSW (M&H) Rules, 2000. The central Solid Waste Management Department should be responsible for procurement and upkeep of vehicles, construction of transfer stations, setting up and maintenance of processing plants as well as for managing the disposal sites in an environmentally acceptable manner. It should make policy for house to house collection, decide the tax imposition, prepare a module for awareness and public private participation, contract document and suitable terms and conditions for various activities under SWM sector etc.

The subject of solid waste management, proposed to be handled by Environmental engineers/public health engineers or civil engineer of Executive Engineer level with the support of mechanical/automobile engineers to handle the workshop facilities and various processing plants. Qualified engineers should, therefore, be regularly inducted in handling Solid Waste Management.

B. Capacity Building/Training

Capacity building by the following training activities will be essential for effective implementation of the Master Plan for Solid waste Management:

- Training for sanitation workers
- Training for supervisors
- Training for officers
- Training to elected representatives
- Training to other stakeholders
- Mass awareness campaigns for waste generators

Basic Course Content for Training to Various Levels of Staff/Officers

(i) Training for Sanitation workers

- Importance of sanitation in urban areas
- Present scenario of solid waste management system in the urban areas, deficiency in the system etc.
- Impact of inefficient Solid Waste Management services on health and environment.
- Inefficiency of tools and equipments used and loss of manpower productivity.

- Need for modernization of solid waste management practices.
- Options available for improving the system
- Advantages of using improved tools and equipments for primary collection of waste and street sweepings
- Need for synchronization of storage of waste at source, primary collection of waste and waste storage depots
- Proper upkeep of tools and equipments and waste storage depots

(ii) ***Training to Sanitation Supervisors of Various Levels***

- Need for synchronization of transportation of waste with waste storage depots
- Transportation of waste on day to day basis
- Waste processing and disposal options, advantages and disadvantages of various technologies.
- Sanitary land filling and related technologies
- Public and NGO participation in waste management
- Building public awareness
- Enforcement of the various rules

(iii) ***Training for Officers Looking after Solid Waste Management Department***

- Solid Waste Management good practices prevalent in other parts of the country and in the developed countries
- Institutional strengthening, internal capacity building and human resource development
- Private sector participation in Solid Waste Management and cost benefits accrued on account of PSP.

- Management information systems
- Financial, health and legal aspects

(iv) Training to Elected Representatives

Whereas the Municipal Commissioners or the Chief Executives are responsible for day-to-day affairs of the urban local bodies, the elected members are the policy makers and their sanctions are essential for any major investments or improvements in Solid Waste Management services. It is, therefore, necessary that the members of the elected wing such as the Mayors of the Corporation and other important office bearers, are given appropriates orientation towards the need of modernization of solid waste management practices in the urban areas and the importance of the same in terms of health and sanitation in the cities/towns.

(v) Mass Awareness Campaigns

For effective implementation of SWM practices, multi pronged action is required in association with all stake holders- Municipal corporation employees, Public representatives, State Government, NGOs and waste generators.

- On one side, training and refresher courses are essential for Municipal employees; the other important side is conducting mass awareness programmes for waste generators.
- People need to know about the duties of municipality/municipal corporation; rules and regulations/obligatory functions; as well as their participatory role towards city's cleanliness. Information, Education and Communication (IEC) mechanisms should be used to ensure effective public participation
- Following methodology is suggested for carrying out **community awareness** and **public participation** in the sector of Solid Waste Management:

- Dissemination of information to community regarding best practices of Solid Waste management through print, electronic media and unconventional media.
- Design and implement a suitable mechanism for source segregation, collection, transportation and proper disposal on pilot basis through NGO.
- Capacity building of different stakeholders involved viz. Non Government Organisations (NGOs), Community Based Organisations (CBOs), Volunteers, line agencies, etc. for development of proper mechanism for adopting best practices of solid waste management
- Identification of major issues of concern in respective area-such as,

  i. Need and urgency of waste disposal

  ii. Obligatory functions of Urban Local Bodies (ULB) and legal provisions for people's participation

  iii. Present scenario of SWM and need for active role of community participation

  iv. Need of formation of CBOs, Neighborhood committees (NCs) for timely disposal and collection of waste; recycling, reusing and reduction of waste at source; segregated storage at source; deposition of waste at proper place in proper containers; no littering on streets and public places; paying user charges regularly and cooperation to Municipal corporation employees

  v. Role of N.G.Os in city/Mohalla/ward cleanliness drive, street sweeping

  vi. organizing awareness camps, skill enhancement of the people/communities as well as develop coordination with the ULB

- Awareness generation through electronic media:

- i. Scroll on message
- ii. Ad spots
- iii. Songs with drama
- iv. Interactive sessions
- v. Quiz competitions

➢ Awareness generation through print media:

- i. Appeals
- ii. Advertisements
- iii. Articles
- iv. Distribution of pamphlets
- v. Stickers
- vi. Posters

➢ Awareness generation through unconventional and traditional media:

- i. Street plays
- ii. Advertisement through government buses/private buses
- iii. Organization of workshops on 'best out of waste' during hobby classes and in schools
- iv. Organization of public meetings, door to door campaigns etc.
- v. Ensure facilitating awareness generation campaign in a series.
- vi. Facilitate having associating partners, like NGOs, line agencies, media to implement joint venture for awareness generation campaign
- vii. Capacity building of municipal personnel, NGOs, CBOs for development of effective mechanism for adopting best practices

viii. Formation of committees at local level to sustain the initiative undertaken and incorporating related capacity building initiatives

Don't Waste the Waste

As Delhi goes crazy on construction ahead of the Commonwealth Games, the civic body has tied up with a firm to recycle tones of building and demolition waste. 4,500 metric tones of construction waste dug up in Delhi every day. 80 per cent of that which account for waste is recycled. This the first of its kind recycling plant, possessing construction material, in the country. The plant started functioning in Burari (west Delhi) in January, 2010. Construction and demolition waste is collected from 42 collection points identified by the Municipal Corporation of Delhi (MCD) and transported to the plant site. Recyclables are separated from the waste before they can reach the crushing stage. The waste received is then crushed using crushers. Recycled material is used for land filling, laying of roads, pavement construction, etc.

Conclusion

Solid waste management is a large public service system, which needs to be efficiently provided to the community to maintain aesthetic and public health standards. The scientific way of collection, transportation, processing and disposal, financial provisions, public participation and awareness will help in making the city environment neat and clean and will help in long run to reduce the cost of its inhabitants on health/ medical treatment. There is also need of a strong will and dedication to implement the rules and laws relating to solid waste management.

References

Bandela, N.N. and Tare, D.G. (2009), Municipal Solid Waste Management, B. R. Publishing Corporation, Delhi.

Central Pollution Control Board (CPCB), (2000), "Management of Municipal Solid Waste" CPCB,New Delhi

Goyal, S.K., (2008), "Toolkits - Master Plan Preparation". Unpublished document, presented in a Training Programme organized by

Association of Municipalities and Development Authorities (ASMDA), New Delhi on 14 September, 2008.

Hindustan Times, (2010), "MCD: No Wasting Waste", Hindustan Times New Delhi edition,1st March,.

Kumar, Sunil (2005), "Municipal Solid Waste Management in India: Present Practices and Future Challenge", Asian Development Bank, Manila, August.

Ministry of Environment and Forest (MoEF), (2000), "Municipal Solid Waste (Management and Handling) Rules", MoEF ,New Delhi.

Sharholy, M., Ahmad, K., Mahmood, G., and Trivedi, R.C.(2008), "Municipal Solid Waste Management in Indian Cities - A Review", Journal of Waste Management, Volume 28, Issue 2..

Singhal, Shaleen and Pandey, Suneel, (2001), "Solid Waste Management in India: Status and Future Directions", TIMES, Volume 6, Number 1, New Delhi.

WilberSmith, (2009), "Solid Waste Management Master Plan of Ghaziabad, Capacity Development of the National Capital Region Planning Board (NCRPB)" - Component B, Interim Report, Volume II D, 2009, NCR Planning Board and Asian Development Bank, New Delhi.

www.ec.europa.eu

www.jnnurm.com

Chapter 8

Pricing Solid Waste for Effective Management

Krishan K Yadav

Economic development is a broad concept that includes economic growth as well as welfare of masses. A long term rise in income leads to economic growth. National policies intended at higher growth rate generally vie for the ways & means of achieving their goals through increased economic activities - production, distribution, consumption, trade are some of the elements involved in these economic activities. While on the one hand enhancement of these activities leads to generation of higher income, it simultaneously may pose serious challenges such as pollution, garbage, global warming, environmental degradation etc, to human welfare. So there are often the cases where efforts towards growth hamper the development. Of such many concerns, the challenge of garbage disposal is a major issue. The daily collection Municipal Solid Waste (MSW) is rising year by year at a rate higher than the population growth. This is evident from the fact that during the last census decade while the population of Mumbai grew from 8.2 million to 12.3 million i.e. a growth of 49 percent, the MSW of the city increased at a higher rate of around 67 percent. So it is not only the population that can be held responsible, the problem somehow has linkages with other factors such as consumption and changing living styles that in turn are outcomes of income. So how far sensible would it be if a minor part of the income of the consumer responsible for unnecessary creation of solid waste is diverted towards its management, is the subject matter of the present paper. The paper initially highlights some of the issues relating to generation of solid waste and its ill effects of various kinds then suggests an economic policy for efficient management of solid waste generated from the domestic sources.

Waste and Disposal

Disposal means any operation which may lead to resource recovery, recycling, reclamation, direct re-use or alternative uses. In ordinary terms waste means a substance which is not needed by the owner, producer or processor. In technical terms however, waste refers to substances or objects which are disposed of or are intended to be disposed of or are required to be disposed of by the provisions of the law. These wastes can be classified in terms of their state such as solid or liquid waste and also in terms of their properties such as bio-degradable and no-biodegradable waste. Of various solid wastes, which is the matter of discussion in the present paper substances such as paper, wood, fruits, vegetables, leaves and others fall into bio-degradable category while plastics, metals, foams, thermo-coals, glass do not bio-degrade. Among various solid wastes some substances are hazardous while some others are not much hazardous.

Sources of Solid Waste

All solid wastes emerge from two main sources - domestic and commercial. Domestic waste can be

- organic in form of food residual, leaves, vegetables etc.,
- inorganic in form of plastics, metals, papers, woods etc, and;

- the sewage waste.

Commercial waste comes from two main sources viz. from the production units and from the support services. The production waste includes

- sludge,
- inorganic waste such as paint, rubber, plastics etc., and;
- toxic chemicals.

The waste from support services such as canteens, rest houses, housing etc. is similar to the domestic waste - organic, inorganic, and sewage waste.

The problem of waste generation is sure to become alarming, sooner or later. As per the estimates of the Ministry of Environment and Forest, around 7200000 tones of hazardous waste are generated each year and this is rising. The disposal of solid waste requires an additional 1 square kilometre of landfill area every year and the collection, disposal and treatment of solid waste incurs and annual expenditure of around Rs. 1600 crore every year.

Waste Collection in India

This is done by variety of means that include both formal and non-formal means. The major part of the solid waste is collected by the city municipality through *Safai-Karamcharis*, Receptacles and Garbage Collection Vehicles. This waste is transferred to a *Dhalao* from where it is transported by trucks and dumped at the city outskirts. In Delhi the work is undertaken by the Municipal Corporation of Delhi and the New Delhi Municipal Committee. Three dumping sites at Bhalaswa, Gazipur, and Okhla that were earlier located on the city limits have now been surrounded due to urbanization.

The other means of waste collection is in form of *Kabadiwalas, Raddiwalas* and Ragpickers. Their collection includes small iron pieces by magnet, metals, electric wires, waste papers, cardboard carton boxes, glass and plastic bottles, polythene etc. Some of these articles are even purchased from the households' e.g. old newspapers, clothes, shoes etc. They generally sell these articles further for recycling and in a way contribute towards an effective disposal of useful waste. On the contrary in the absence of proper guidance, education and hence the awareness, they can be accused for causing environmental pollution for using unfair disposal methods such as pouring hazardous chemicals into sewers and burning plastics and rubbers to segregate metals. In various unauthorized colonies where these people live large smoke of such burnings can be seen. Lack of awareness drives these people to throw their toxic waste in most dangerous ways.

Ineffective Solid Waste Management

Effects

Unwise waste management has serious adverse effects. It directly affects the health of human beings, environment, climate, marine life as well as the socio-economic condition of the society. Accumulation of solid waste on earth surface causes the air temperature to rise beyond the normal limit. That subsequently raises the sea temperature, which raises the sea levels and change precipitation and other climate conditions. This change in regional climate could alter the forests, crop yields, rains and snowfalls. These factors are sure to affect the ecosystem as a whole. Some of the consequences include expansion of deserts, alteration in the basic feature of jungles and extinction of some of the species. Although, actual effects may be much or less severe, exact determination of which is not possible in advance, still there is likely to be the overall trend towards increase precipitation, increased evaporation, intense rain storms and drier soils. Another serious danger is that of emission of greenhouse gases into the atmosphere.

- Carbon Dioxide (**CO_2**) is released by burning of wood and solid waste having wood products.
- Methane (**CH_4**) is emitted by decomposed materials, compost pits, coal, natural gas, oils sites.

- Nitrous Oxide (**NO_2**) is emitted during agricultural and industrial activities as well as during composition of solid waste and fossil fuels.

The above discussion is a clear indication of the ill effects that may occur due to improper disposal of solid waste. Some accidental happenings during recent years also support the argument.

- During 2005, a cloud burst occurred in Mumbai because of choked sewer line due to large quantity of waste polythene bags.
- A blast occurred in the imported scrap from Iran in Bhushan Steel Factory in NOIDA during 2007.
- Contaminated food and changed climate conditions have caused significant reduction in number of migrant birds.

Causes

The prime cause of generation of domestic waste can be associated with rising consumption that itself is an outcome of rising incomes. The income rise can again be associated with higher economic growth, the prime link of which is with the long term rise in income. So generation of this additional waste is an outcome of economic growth that will become development when this waste is disposed off effectively, preventing the humans of the ill effects.

India has come a long way from its inward-looking economic strategy of over 50 years ago. Economic liberalization and the gradual opening up to the world have boosted growth and. India has become one of the fastest growing economies out of 34 developed and emerging markets and it is expected to become one of the largest 2020. The implications of the growth trajectory are manifold. One of these is the emergence of a larger and richer consumer market, and second, the change in the consumption patterns, with expenditure on consumer goods growing substantially. Third, household savings have increased, giving the sense of security to large number of people. Fourth, there is rising demand for diversified financial instruments to invest those savings. This is totally against the investment patterns of the households that have characteristically been conservative and poorly diversified.

Given India's relatively young population, a lower risk aversion leading to increasing demand for consumer goods can be expected. Adding to that a rising income level for this population group, and a defined-contribution pension reform in the making, the scope for expansion appears large going forward.

The rising income has brought about changes in lifestyle. With such advancements in living standards, the trend of reusable articles used in long run is fast disappearing. As again earlier time the tendency of repairs has shifted towards substitution. This statement can be testified by simple market observations. The mechanics and repair persons present a couple of decades ago have now given way to sellers. To find a television, radio and watch mechanics is not a normal task now while during the eighties they would flood the market and work in a competitive manner but still would earn plenty as there was equally large number of customers. Insufficient work has forced them to change their job. Discarding of such articles creates a lot of solid waste.

A look into a normal house and its virtual comparison with the house of same standard of the past also reveals certain interesting points. The normal household goods have risen a great deal owing to new inventions and their consequently affordable price. Some such goods do not last very long and once retired, have no second-hand value. Computer peripherals and add-ons - printers, scanners, disk drives, and compact disks are not that long lasting. When out of order they are cheaper to replace rather than repair and are totally useless thereafter, create a lot of solid waste of non-biodegradable category.

Pricing Solid Waste

In order to reduce the quantity of solid waste it is mandatory that less and less quantity is generated and even if generated, most part of it is recycled at source. Two approached that can be adopted here are of encouragement and enforcement nature. Encouragement through education and awareness campaigns is essential. However, in case of certain large cities where the problem is getting beyond the limits; enforcement may also be applied in form of pricing of solid waste generated. This price is to be paid by the person generating the garbage to the disposal authorities. Quantity based pricing of the solid waste on one hand will result in less garbage and on the other hand reduces the financial burden of the disposal authorities.

The assumption that rising garbage heaps are resultant of rising incomes does not appear baseless. The situation under which the rise in quantity of solid waste is higher than the corresponding rise in population is not at all favourable. Per households solid waste generation now is more than what it was some twenty years ago. The need of the time is to control solid waste at its emergence. Transferring a minor part of the income of the persons responsible for waste generation to the disposal authorities should work in multiple ways in the direction of solid waste reduction. The scope of the present paper has been limited to pricing of domestic waste; the next few paragraphs discuss the issue in details.

Pricing the solid waste is one method that could be effective in reducing overall waste generation and conserving economic and environmental resources. Under solid waste pricing, household consumers are charged for waste collection and disposal services based on the amount of trash they generate. The theory is that if consumers know they must pay more to produce more garbage they will take advantage of source reduction and recycling opportunities to reduce their trash-and their trash collection bills as with other charged utilities - electricity, water, cooking gas etc. consumers will limit usage if their actions directly affect the cost of their services. The system is already working successfully in US, Canada and some other western countries.

Pricing Methods

Under solid waste pricing method for domestic households' consumers pay for their waste generation. Two methods can be adopted - one the unit price per bagful of garbage method, and; two, the price per unit weight of garbage method.

Under the first method i.e. the unit price per bag the initial price is fixed at a minimum level. For example those generating up to two bags of garbage per week pay the same amount as those who generate one third of that quantity. Here the consumers are charged a fixed amount up to a certain level (two bags in above illustration). If the quantity exceeds that level, the consumers are charged at variably increasing rates. Say, as for the above illustration Rs. 5 up to two bags, additional Re. 1 for the third bag, and additional Rs. 2 for the fourth bag and so on and so forth. Under such solid waste pricing set up consumers are charged for what actually they throw away, creating an incentive to minimize household waste generation and recycle the compost.

Another method could be weight based method. This method is expected to be rather difficult to adopt and maintain and also is likely to take more time per household for waste collection. Although weight based system may provide even a greater incentive to reduce waste as the citizens must reduce their trash by an entire bag to see a difference in their structure, it may be technically more complicated to implement. It would require retrofitting garbage collection vehicles with scales and electronic bar-coding equipment to enable the authorities to charge the consumers by kilograms. Under this method, the consumers have little chance for adjustment if the quantity of garbage exceeds the maximum limit for the week.

The trash bag used to collect the garbage must bear an official tag and should be supplied preferably by the authorities either free of cost or at a price to minimize the waste. These trash bags for bio-degradable and non-biodegradable waste should bear different colour codes and also a mention of the household disposing off the waste to ensure proper segregation.

Complicated or simpler, the choice should be decided keeping in mind various local, environmental, social and economic conditions. Whichever of the two methods is adopted, the key to its effectiveness is establishing a link between a consumer's solid waste disposal practices and the fee charged for solid waste management service. This link provides the incentives for consumers to reduce their waste through source reduction and recycling.

Pricing Issues

Wide diversity exists between various communities in terms of size, location, economic resources, available vacant land and population. Waste management needs also differ widely from community to community. Whether or not to implement the pricing method and if yes, which method is to be adopted is a question that needs to be satisfied adequately before taking the final decision. Various issues are needed to be examined in this regard.

Issues common to all sorts of communities and also common to both pricing and conventional methods are:

- Who will provide the service?
- How often will the waste be collected?
- Will the service be backyard collection or curb side pick?

The issues, considered unique to pricing of solid waste include

Community Support

The design of the programme must be such that it is appealing and convenient for the consumers. This must be ensured at the outset that rates are perceived as fair, alternative to disposal such as recycling etc., must be available and cost effective. A big support in this direction could be through distribution of brochures and campaign through media.

Enforcing and Monitoring

In Indian context, among a large part of residents, there has been a tendency to avoid the payment burden of public services by use of unfair means. Tempering electricity and water meters to reduce the bill amount is a common unfair practice in most Indian cities as well as in villages. The chance that such illegal practices could extend to solid waste if it is priced is not null and void. Moreover, the practices of littering, backyard burning, dumping into common public receptacles and sewage are easier to do in comparison with electricity and water meter tempering. Such problems may also be needed to be addressed by the waste managers. The authorities, therefore, should have a strategy to combat such waste diversion tactics. Proper enforcement and compliance monitoring like imposition of ban and stiff fines can significantly reduce the problem.

Determination of Rate

The determination of solid waste pricing rate depends upon the objectives of the authorities responsible for solid waste disposal i.e. whether it is intended to fully finance the disposal cost by pricing or the intentions are aimed at encouraging the community to reduced and recycle their waste. In first case however, the pricing should be such as to ensure collection of minimum support revenue even if the households produce the minimum solid waste. Another decision in this respect may depend upon the littering and garbage producing habits of the residents. At higher price, consumers will probably waste less and recycle more.

The System Design

For the system to be viable, it has to be so designed that residents feel comfortable about it and the authorities are also at ease. The time of collection, design of collection vehicles, methods and timings of transportation are all very important. Other factors in this respect include availability of funds for implementation, community cooperation, flexibility of revenue requirement etc. Also there could be situations where the waste managers may need to develop alternative rate options. For the cases of apartments and multi family house, who use to dump their waste at a common place, present a complex problem. Here no single family would take the responsibility to pay for the garbage. In such a case it should be mandatory for each family to throw their garbage in their individual trash bags. Non-compliance should be fined. There should be variable rate options also - the low-income families and backward communities such as *jhuggi jhopri* (slum) cluster should be charged at differential rates.

Support Measures

The rising garbage heaps poses a serious threat to the environment and the health of the living beings. The success of garbage pricing programme is likely to be more effective when it is implemented in conjunction with recycling and composting programmes. People can easily find various recyclables in their trash. Items such as yard waste, glass bottles, tin, aluminium, newsprints, cardboard, corrugated sheets, plastics, paper are all recyclable. Segregation of these items will drastically reduce the quantity of waste requiring conventional waste. The authorities however are required to provide ample recycling facilities to the residents. Separate, distinct and adequate receptacles for such segregated articles should be available at convenient locations for the households.

Table 1: Composition of Municipal Solid Waste in Delhi

Description	Percentage of Composition by Weight
Vegetable residual and leaves	40.15
Grass	3.80
Paper	0.81
Plastic	0.62
Glass and ceramics	0.44
Metal	0.64
Stone and ash	41.84
Miscellaneous	11.73

Source: Central Pollution Control Board, Delhi

About 45 percent of the total solid waste is bio-degradable. Only a little proportion of this is converted into compost presently. The plants are not functioning at their capacity for many reasons. Pricing of domestic waste at Re 1 per 10 kg can be expected to earn about Rs 7 crore per year even at underestimations. The collection from commercial waste would be higher. Imposition of pricing system is likely to lower the quantity of solid waste provided there are support facilities for alternative disposal at source. With the people having access to multiple recycling and reduction options, the results will be favourable. The programme is likely to be more successful in educated and advance communities who give more thought to such sensitive issues. The backward and poor communities are needed to be effectively educated in this regard. The consumers may also be encouraged to reduce garbage by using fewer packaged goods and throw away items. This practice will save them money, encouraging further in this direction. The pricing method of the solid waste residual is likely to be more cost effective in longer run as the cost of service is bound to decrease with reduction in quantity of trash and participation of community. The programme is likely to be viewed as clearer as it increases savings, reduces garbage and acts in the direction of betterment of the society as a whole.

References

US Environmental Protection Agency(USEPA) 1991, *"Variable Rates in Solid Waste: Handbook for Solid Waste Officials"*, US Environmental Protection Agency(USEPA) 1991, Washington DC.

US Environmental Protection Agency(1995), *"The Effects of Weight- or Volume-Based Pricing on Solid Waste Management"*, US Environmental Protection Agency(USEPA) 1991, Washington DC

Jennifer Asuncion Mund (2005), *India Rising – A Medium Term Perspective, Deutsche Bank Research.*

Hermann Lotze-Campen, Katrin Gerlinger, Fritz Reusswig (2003), *"Changing Global Lifestyle and Consumption Patterns:The Case of Energy and Food"*, Potsdam Institute for Climate Impact Research (PIK), Global Change & Social Systems Department.

Rajaram Vasudevan, Mr. George Mathew, "*Waste Management - Indian Scenario*", Techno Engineering, Chicago, IL, USA and W2ES USA Inc, Cochin, India.

Sanjeev Agrawal, "*Municipal Solid Waste Management in Urban Centres*", Central Pollution Control Board, Delhi

Shaleen Singhal and Suneel Pandey, (2001), "*Solid Waste Management in India: Status and Future Directions*", The Energy and Resources Institute, New Delhi.

Shekdar (1999), "*Municipal Solid Waste Management – the Indian Perspective*", Journal of Indian Association for Environmental Management, 26 (2): 100–108.

Chapter 9

Hazardous E-Waste: Callenges and Remedies

Mridushi Swarup and Ritika Banerjee

Introduction

Technology designed to make life easier and more comfortable. Along with economic growth, the high tech boom and fast dipping cost of technology has allowed people the luxury of buying the new electronics instead of repairing or upgrading their existing item. This convenience comes at a cost: hazardous pollution in the form of electronic waste (e-waste). E-waste is one of the rapidly growing environmental problems in India.[1] Although electronic technology has contributed to immense advances in modern society,[2] it has also created a rapidly growing toxic waste stream.[3] Electronic waste, commonly referred to as "e-waste," is one of the most lethal and fastest growing sources of municipal waste. Technology is developing rapidly, products are becoming cheaper, and as consumers purchase the newest and trendiest items, they are discarding their older products in record numbers.[4] Consumers hardly ever ask important questions, such as where their waste goes, what chemicals are in their waste, and whether these chemicals will harm other human beings or the environment. This is tragic not only because e-waste contains dangerous chemicals, the improper handling of which harms people and the environment, but also because these harms are preventable.[5]

Considering the current technological scenario, it is very essential to fathom the importance of issues as grave as e-waste and its management. Against this background, the present paper deals with different aspects of electronic waste..

The Meaning of E-Waste

The term "e-waste" is loosely applied to consumer and business electronic equipment that is near or at the end of its useful life. There is no clear definition for e-waste. However, it can be defined as any and all electronic appliances that are discarded because of malfunction, exhaustion or obsolescence.[6] E-waste is a popular, informal name for electronic products nearing the end of their "useful life."[7] In other words, electronic waste or e-waste is the term used to describe old, end-of-life electronic appliances such as computers, laptops, etc. which have been disposed of by their original users.[8] As there is no generally accepted definition of e-waste, in most cases, e-waste comprises of relatively expensive and essentially durable products used for data processing, telecommunications or entertainment in private households and businesses.[9]

Computers, televisions, VCRs, stereos, copiers, and fax machines are common electronic products. Many of these products can be reused, refurbished, or recycled. Unfortunately, electronic discards is one of the fastest growing segments of our nation's waste stream.[10] In addition, some researchers estimate that nearly 75 percent of old electronics are in storage, in part because of the uncertainty of how to manage the materials. Combining this with increasing advances in technology and new products headed towards the market, it is no wonder that "e-waste" is a current and widespread issue.[11] Public perception of e-waste is often restricted to a narrower sense, comprising mainly of end-of-life information & telecommunication equipment and consumer electronics. However, technically, electronic waste is only a subset of WEEE (Waste Electrical and Electronic Equipment). According to the OECD any appliance using an electric power supply that has reached its end-of-life would come under WEEE.[12]

While e-waste contains both valuable materials, it also contains harmful substances. In the absence of suitable techniques and protective measures, recycling e-waste can result in toxic emissions to the air, water and soil and pose a serious health and environmental hazard.[13]

E-waste per se can be divided into numerous categories[14]. Some of these categories include:

- Obsolete electronic devices like computers, servers, main frames, monitors, TVs & display devices.[15]
- Large household appliances and Small household appliances like refrigerators, air conditioners, washing machines, and microwave ovens.[16]
- IT and telecommunications equipment such as cellular phones & pagers, calculators, audio and video devices, printers, scanners, copiers and fax machines.[17]
- Consumer and Lighting Equipment and Monitoring and control instruments
- Electrical and electronic tools (with the exception of large-scale stationary industrial tools) such as industrial electronics which includes sensors, alarms, sirens, security devices, automobile electronic devices and electronic components such as chips, processors, mother boards, printed circuit boards.[18]
- Toys, leisure and sports equipment like DVDs, CDs, floppies, tapes, printing cartridges[19]
- Medical devices with the exception of all implanted and infected products
- Automatic dispensers[20]

Toxic Constituents of E-Waste

E-waste is not hazardous *per se*. However, the hazardous constituents present in the e-waste render it hazardous when such wastes are dismantled and processed, since it is only at this stage that they pose hazard to health and environment.[21] All electronic items contain thousands of substances including lead and mercury, which are very dangerous to health.[22] The harmful materials contained in electronics products, coupled with the fast rate at which the outdated units are getting replaced, pose a real danger to human health and the environment if electronics products are not properly processed prior to disposal.[23] The composition of e-scrap depends strongly on the type and the age of the scrap.[24] E-waste contains considerable quantities of valuable materials such as precious metals also.[25]

Electronics products like computers and cell phones contain a lot of different toxins. Up to 38 separate chemical elements are incorporated into electronic waste items[26].Hazardous materials in one form or the other are present in such wastes primarily consisting of Cathode Ray Tubes (CRTs), Printed board assemblies, Capacitors, Mercury switches and relays, Batteries, Liquid Crystal Displays (LCDs), Cartridges from photocopying machines, Selenium drums (photocopier) and Electrolytes.[27] For example, CRTs of computer monitors contain heavy metals, which can be very harmful to health.[28] All electronic equipments contain printed circuit boards which are hazardous because of their content of lead, brominated flame retardants and antimony oxide, which is also present as a flame retardant[29].

Table I : Toxic Constituents Of E-Waste

COMPONENTS	CONSTITUENTS
1. Printed Circuit Boards	Lead and Cadmium
2. Cathode Ray Tubes (CRTs)	Lead Oxide, Cadmium and Baruim
3. Switches and Flat Screen Monitors	Mercury
4. Computer Batteries	Cadmium
5. Printed Circuit Boards, Plastic Casings and Cable	Brominated Flame-Retardant Casing
6. Cable Insulation/ Coating	Polyvinyl Chloride (PVC)
7. Motherboards	Beryllium
1. Plastic Housing	Bromine[30]
2. Cooling Unit, Insulation Foam	Cholorofluorocarbon (CFC)
3. Light Emitting Diodes	Arsenic
4. RechargeableNiCd-batteries, fluorescent layer (CRT screens), printer inks and toners, photocopying-machines (printer drums)	Cadmium[31]
5. Rechargeable NiCd-batteries or NiMH-batteries, electron gun in CRT	Nickel[32]
6. Fluorescent layer (CRT-screen)	Rare Earth elements (Yttrium, Europium)
7. Older photocopying-machines (photo drums)	Selenium
8. Medical equipment, fire detectors, active sensing element in smoke detectors	Americium[33]

According to a United Nations report, developing countries face increasing environmental and health hazards from electronic waste unless toxic materials are collected and recycled properly.[34] According to report, titled Recycling - from E-Waste to Resources, the world produces about 40 million tons of waste from electronic devices, known as e-waste, every year.[35]

Generation of E-Waste

The lightning speed of growth of the Information Technology (and now mobile telecom) industry is leading to fast-growing piles of e-waste worldwide, particularly in developed nations, and unless they are quickly and safely disposed off, e-waste is going to take its huge, fearsome and absolutely irreversible toll on all life and the environment soon.[36]

Global economics and the digital divide encourage the export of hazardous e-waste to poor, developing countries that are far less able to manage the environmental effects of e-waste.[37] The economic logic behind dumping a load of toxic waste in the lowest wage country is impeccable.[38] The export of toxic e-waste from rich, developed countries to poor, developing countries is a consequence of global economic forces[39] because hazardous wastes generally follow the path of least resistance—that of lower costs and lower standards.[40] The economic incentives created by strict domestic processing and environmental regulation, lax or nonexistent regulations in developing countries, and the ease of free trade brought about by globalization, force even well-intentioned recyclers to export e-waste.[41]The e-waste stream is unlike other hazardous waste streams because its valuable components and capacity to be reused or recycled can potentially provide economic incentives for efficient waste management.[42] It is estimated that 50 percent of computers turned in for recycling are in good working order[43]. Further, many components from nonworking electronics could simply be transferred to new machines[44]or recycled for their valuable raw materials.[45] However, much of the potential for reusing e-waste is not realized because electronics are designed for disposal rather than recycling, making them expensive and difficult to disassemble or upgrade.[46]In addition, as manufacturers work to cut escalating production costs, the precious metal content in electronic products is reduced,[47]further reducing the economic incentive to recycle electronics domestically. Therefore, without regulations making private actors internalize the negative externalities of e- waste, potential incentives for domestic reuse and recycling are not strong enough to overcome the greater economic pressures to export.[48]

The first comprehensive study to estimate the annual generation of e-waste in India and answer the questions above is being undertaken up by the National WEEE Taskforce. So far the preliminary estimates suggest that total WEEE generation in India is approximately 1,46,000 tonnes per year. The top states in order of highest contribution to WEEE include Maharastra, Andhra Pradesh, Tamil Nadu, Uttar Pradesh, West Bengal, Delhi, Karnataka, Gujarat, Madhya Pradesh and Punjab. The city wise ranking of largest WEEE generators is Mumbai, Delhi, Bangalore, Chennai, Kolkata, Ahmedabad, Hyderabad, Pune, Surat and Nagpur. India generated 3.3 lakh tonnes e-waste in 2007 and is expected to touch 4.7 lakh tonnes by 2011, citing the press release of the inventory study of MAIT and GTZ presented in December 2007.[49]

Impact of E-waste

The harmful materials contained in electronics products, coupled with the fast rate at which we're replacing outdated units, pose a real danger to human health if electronics products are not properly processed prior to disposal.[50]

Impact of Informal Recycling

Electrical and electronic equipment are made up of a multitude of components, some containing toxic substances which can have an adverse impact on human health and the environment if not handled properly. Often, these hazards arise due to the improper recycling and disposal processes used. [51]The accrued electronic and electric waste in India is dismantled and sorted manually to fractions such as printed wiring boards, cathode ray tubes (CRT), cables, plastics, metals, condensers and other, nowadays invaluable materials like batteries. It is a livelihood for unorganised recyclers and due to lack of awareness, they are risking their health and the environment as well.[52] The valuable fractions are processed to directly reusable components and to secondary raw materials in a variety of refining and conditioning processes. No sophisticated machinery or personal protective equipment is used for the extraction of different materials.[53]

Effect on Health

Electronics like computers and cell phones contain a lot of different toxins. For example, cathode ray tubes (CRTs) in computers contain heavy metals, such as lead, barium and cadmium, which can be very harmful to health if they are recycled without taking proper precautionary measures and if they enter the water system.[54] These materials can cause damage to human nervous and respiratory systems. Flame retardant plastics, used in electronics casings, can release particles that damage human endocrine functions. These are the types of things that can happen when unprocessed e-waste is put directly in landfill.[55] E-waste contains a brew of toxic substances, such as lead (responsible for causing damage to the central and peripheral nervous system, circulatory system, kidney, reproductive system, and endocrine system, as well as slowing brain development in children)[56], cadmium (widely believed to cause irreversible effects on human health upon accumulation in the human body, particularly the kidneys)[57], mercury (responsible for causing damage to various organs including the brain and kidneys, as well as to fetuses)[58], chromium-VI (causing toxic effects in the cells and damage to DNA)[59], plastics[60] (including PVC)[61], Brominated Flame Retardants (BFRs)[62], and beryllium (which can cause lung cancer and beryllicosis).[63]

Table 2 summarizes the health effects of certain constituents in e-wastes. If these electronic items are discarded with other household garbage, the toxics pose a threat to both health and vital components of the ecosystem.

Table 2: Effects of E-Waste Constituent on Health

Source of E-Wastes	Constituent	Health Effects
Solder in printed circuit boards, Lead (PB) glass panels and gaskets in computer monitors		• Damage to central and peripheral nervous systems, blood systems and kidney damage.[64] · Affects brain development of children. [65]

Chip resistors and semiconductors	Cadmium (CD)	• Toxic irreversible effects on human health. · Accumulates in kidney and liver.[66] • Causes neural damage. • Teratogenic.
Relays and switches, printed circuit boards	Mercury (Hg)	• Chronic damage to the brain. • Respiratory and skin disorders due to bioaccumulation in fishes. [67]
Corrosion protection of untreated and galvanized steel plates, decorator or hardner for steel housings	Hexavalent chromium (Cr) VI	• Asthmatic bronchitis • ·DNA damage.
Cabling and computer housing	Plastics including PVC	• Burning produces dioxin. It causes • Reproductive and developmental problems; · Immune system damage; · Interfere with regulatory hormones[68]
Plastic housing of electronic equipments and circuit boards.	Brominated flame retardants (BFR)	• Disrupts endocrine system functions • Causes cancer[69]
Front panel of CRTs	Barium (Ba)	Short term exposur e causes: · • Muscle weakness; • Damage to heart, liver and spleen.
Motherboard	Beryllium (Be)	• Carcinogenic (lung cancer) · Inhalation of fumes and dust. Causes chronic beryllium disease or beryllicosis. • Skin diseases such as warts. [70]

Effect on Environment

Electronic waste also invariably harms the environment. Computers and other electronics include toxic substances [71] and precious metals.[72] Usually, the precious metals are extracted from the electronic goods, and the remainder of the dismantled products is burned or discarded in landfills.[73] Lead, mercury and other heavy metals found in electronics pose environmental and public health risks because they can contaminate groundwater when released into the environment.[74] Improper disposal and export to rural communities place citizens, both makeshift recyclers and uninvolved persons, at grave risk of health and environmental problems, according to Greenpeace International.[75] When these components are illegally disposed and crushed in landfills, the lead is released into the environment. Lead accumulates in the environment, and has highly acute and chronic toxic effects on plants, animals and micro organisms.[76]

Open air burning of the flame-retardant plastics, often found in electronic products, releases harmful heavy metals, dioxins, and beryllium into the environment [77] and used acid liquids are often dumped into nearby water bodies.[78]The toxic fall-out from open air burning affects both the local environment and broader global air currents, depositing highly toxic byproducts in many places throughout the world.[79] Landfilling of e-wastes can lead to the leaching of lead into the ground water. If a CRT is crushed and burned, it emits toxic fumes into the air. These products contain several rechargeable battery types, all of which contain toxic substances that can contaminate the environment when burned in incinerators or disposed of in landfills.[80] Therefore the extent of environmental risks and health hazards which is caused due to the escalating generation of e-waste is a matter of serious concern.

International Scenario

Some of the biggest brands in electronics industries and their suppliers are contaminating rivers and underground wells with a wide range of hazardous chemicals, according to a Greenpeace survey. An analysis of samples taken from industrial estates in China, Mexico, the Philippines and Thailand has revealed the release of hazardous chemicals during the manufacture of printed wiring board (PWB) and semiconductor chips, and the component assembly. The highly toxic polybrominated diphenyl ethers (PBDEs), widely used as brominated flame-retardants, are used in a wide range of processes.[81]

To deal with the above problems, the various agreements signed and legislations enacted by the countries across the world include:

The Basel Convention

Basel Convention on the Control for the Trans boundary Movement of Hazardous Wastes and Their Disposal[82], in force since 1992, is the only global environment agreement on waste and now comprises almost 165 national governments (April 2005). The global agreement regulates trade in hazardous wastes, including WEEE; it seeks to ensure that they are disposed of safely and that the generation of such wastes is minimized.[83] An Amendment to the Convention, commonly known as the Basel Ban[84], which calls for prohibiting the export of hazardous waste from OECD to non-OECD countries, is still to come into force. India ratified the convention in 1990, however the Basel Ban has still to be accepted.[85]

Findings Stated In Report by BAN (Basel Action Network)

- 50 to 80% E-wastes collected are exported for recycling by U.S. Export is legal in U.S.
- Export is due to cheaper labour and laxed standard in poor countries.
- E-waste recycling and disposal in China, India and Pakistan are highly polluting.
- China has banned import of E-waste.
- Lack of responsibility on the part of Federal Government and Electronics Industry,

Consumers, recyclers and local governments towards viable and sustainable options for disposal of E-wastes.[86]

European Union

The European Union (EU) has enacted two Directives, the first on product recycling (Waste Electrical and Electronic Equipment, WEEE) and the second limiting the use of certain substances (Restriction on the use of Hazardous Substances, RoHS). By August 2005, these Directives were to be implemented in the national laws of each EU member State. The WEEE Directive requires EU member states to set up E-waste collection systems so that consumers may dispose of their electronic goods free of charge.[87] The RoHS Directive aims at reducing hazardous E-waste at the producer level.[88]

Ordinance on the Return, the Taking Back and the Disposal of Electrical and Electronic Equipment (ORDEE)

The Swiss[89] Ordinance on the Return, the taking back and the Disposal of Electrical and Electronic equipment (ORDEE) entered into force on 1 July 1998 and was worldwide the first legal regulation on WEEE.[90] Under the ORDEE, retailers, manufacturers and importers are required to take back, at no charge, appliances of the kind that they normally stock. Consumers, for their part, are obliged to return end-of-life appliances and are not allowed to dispose of them via household waste or bulky item collections.

German Electrical And Electronic Equipment Act

Entered into force in March 2005 and is governing the Sale, Return and Environmentally Sound Disposal of Electrical and Electronic Equipment and implements the EU directives on WEEE and RoHS. The goal of this act is to prevent waste from electrical and electronic equipment, to reduce waste volumes through reuse, provisions for collection, recovery and recycling quotas and to reduce the content of hazardous substances in equipment.

Indian Scenario

E-Waste Situation in India

Telecommunications and information technology are the fastest growing industries today not only in India but world over. Manufacturers' Association for Information Technology (MAIT) has collected the following statistics on the growth of electronics and IT equipment in India:[91]

1. PC sales were over 7.3 million units during 2007-08, growing by 16 per cent. There is an installed base of over 25 million units.[92]
2. The consumer electronics market is growing at the rate of 13-15 per cent annually. It has an installed base of 120 million TVs.[93]
3. The cellular subscriber base was up by 96.86 per cent during 2007-08. Its installed base is estimated to cross 300 million mark by 2010.[94]

In India, e-waste is mostly generated in large cities like Delhi, Mumbai and Bangalore. In these cities a complex e-waste handling infra-structure has developed mainly based on a long tradition of waste recycling. This is mainly operated by a very entrepreneurial informal sector. Rag pickers and waste dealers found it easy to adapt to the new waste stream, resulting in a large number of new businesses focusing on the re-use of components or extraction of secondary raw materials. So far, the e-waste recycling system is purely market driven.[95]

As there is no separate collection of e-waste in India, there is no clear data on the quantity generated and disposed of each year and the resulting extent of environmental risk.[96] With extensively using computers and electronic equipments and people dumping old electronic goods for new ones, the amount of E-Waste generated has been steadily increasing. At present Bangalore alone generates about 8000 tonnes of computer waste annually and in the absence of proper disposal, they find their way to scrap dealers.[97] India generated 3.3 lakh tonnes e-waste in 2007 and is expected to touch 4.7 lakh tonnes by 2011, citing the press release of the inventory study of MAIT and GTZ presented in December 2007.[98]

The results of a field survey conducted in the Chennai, to assess the average usage and life of the personal computers, television and mobile phone showed that the average household usage of the PC ranges from 0.39 to 1.70 depending on the income class. In the case of TV it varied from1.07 to 1.78 and for mobile phones it varied from 0.88 to 1.70.[99]

The growth rate of the mobile phones (80%) is very high compared to that of PC (20%) and TV (18%). Additionally considerable quantities of e-waste are reported to be imported. However, no confirmed figures available on how substantial are these trans boundary e-waste streams, as most of such trade in e-waste is camouflaged and conducted under the pretext of obtaining 'reusable' equipment or 'donations' from developed nations. The government trade data does not distinguish between imports of new and old computers and peripheral parts and so it is difficult to track what share of imports are used electronic goods.[100]

Actors in Generating E-waste

Importer/Manufacturer: Almost 50% of the PC's sold in India are products from the secondary market and are re-assembled on old components. The remaining market share is covered by multinational manufacturers (30%) and Indian brands (22%)[101]. Besides manufacturers are major contributors of e-waste. The waste consists of defective IC chips, motherboards, CRTs and other peripheral items produced during the production process. It also includes defective PCs under guarantee procured from consumers as replacement items.[102]

Consumer: Individual households: Individual households account for 22% of junk computers in India[103]. The preferred practice to get rid of obsolete computers is to get them in exchange from retailers when purchasing a new computer, or passed on to relatives or friends. Government, public and private sector: The business sector accounts for 78% of all installed PC's in India. Obsolete computers from the business sector are sold by auctions. Sometimes educational institutes or charitable institutions receive old computers for reuse. Import of e-waste is legally prohibited. Nevertheless, there are reports of e-waste imports from abroad. Analyses by the ministry of environment showed no results concerning import of e-waste, but the ministry admits that a 100% control of the borders is not possible.[104]

***Traders / Scrap dealers / disassemblers*:** One single person can work as a a trader, scrap dealer as well as disassembler. Some of these traders carry out, for example, the primary work of disassembling obsolete computers, they reuse the working components, assemble new computers and sell them again on the secondary market. Non working components are sold to recyclers.[105]

***Recyclers*:** The market for e-waste in Delhi is not concentrated in a single place, but spread over different areas, each handling a different aspect of recycling. No sophisticated machinery or personal protective equipment is used for the extraction of different materials. All the work is done by bare hands and only with the help of hammers and screwdrivers. Children and women are routinely involved in the operations.[106]

Disposal: Computer waste, which does not have any resale or reuse value, is openly burnt or disposed off in landfills.[107]

Status of E-Waste Management in India

A national framework is required as a basis for the implementation and more importantly for the replication of Clean e-Waste Channels. This framework is developed in a joint effort of all relevant stakeholders. The cooperation platform, led by the Ministry of Environment (MoEF), is the National e-waste Strategy Group which is subdivided into five committees which are Policy & Legislation, Data & Baseline, Skills & Technologies, Industry Participation & e-Waste Management, Awareness & Campaigning.[108]

Despite a wide range of environmental legislation in India there are no specific laws or guidelines for electronic waste or computer waste.[109] As per the Hazardous Waste Rules (1989), e-waste is not treated as hazardous unless proved to have higher concentration of certain substances. Though PCBs and CRTs would always exceed these parameters, there are several grey areas that need to be addressed. The import of this waste therefore requires specific permission of the Ministry of Environment and Forests.[110]

As the collection and re-cycling of electronic wastes is being done by the informal sector in the country at present, the Government has taken the following action/steps to enhance awareness about environmentally sound management of electronic waste[111]:

- Several Workshops on Electronic Waste Management was organised by the Central Pollution Control Board[112] (CPCB) in collaboration with Toxics Link, CII etc.
- Action has been initiated by CPCB for rapid assessment of the E-Waste generated in major cities of the country.[113]
- A National Working Group has been constituted for formulating a strategy for E-Waste management.
- A comprehensive technical guide on "Environmental Management for Information Technology Industry in India" has been published and circulated widely by the Department of Information Technology (DIT), Ministry of Communication and Information Technology.[114]
- Demonstration projects have also been set up by the DIT at the Indian Telephone Industries for recovery of copper from Printed Circuit Boards.

  Although awareness and readiness for implementing improvements is increasing rapidly, the major obstacles to manage the e wastes safely and effectively remain. These include:

- The lack of reliable data that poses a challenge to policy makers wishing to design an e-waste management strategy and to an industry wishing to make rational investment decisions.

- Only a fraction of the e waste (estimated 10%) finds its way to recyclers due to absence of an efficient take back scheme for consumers.[115]
- The lack of a safe e waste recycling infrastructure in the formal sector and thus reliance on the capacities of the informal sector pose severe risks to the environment and human health.
- The existing e waste recycling systems are purely business-driven that have come about without any government intervention. Any development in these e waste sectors will have to be built on the existing set-up as the waste collection and pre-processing can be handled efficiently by the informal sector, at the same time offer numerous job opportunities.

Initiatives Taken In India

Initiative – 1: Indo-Swiss Pilot Assessment In Delhi: Management, Handling & Practices Of E-Waste Recycling In Delhi

- Appreciation of problem by documenting about e - waste trade in a pilot area.
- Establish e-waste trade value chain to identify stakeholders and methodology for estimation.
- Estimate e-waste quantity and identify existing e-waste treatment & disposal practices.
- Establish e-waste trade economics and identify macro level impacts.
- Future initiatives & action plan[116]

Initiative – 2 : National Level Workshop Organised In Collaboration With Gtz-Asem: 15th March, 2004: Way Forward

- Formation of national level working group on e-waste.

- Policy level intervention and planning for a national level study.
- Planning for a "rapid assessment" in the major WEEE producing regions/cities of India.
- Bringing stakeholders including producers (formal & informal) at a common platform.[117]

Initiative - 3

1. Working Group Formed In July 2004 And Consisting Of:

- MOEF, CPCB, Ministry of IT and regulatory agencies (representatives of SPCBs)
- NGOs and Industry associations/industry (MAIT, TEMA, CETMA, ELCINA, CII representatives).
- Experts in the field and producers both formal & informal mandate: identify, plan and implement all issues related to e-waste in India two meetings of working group held in 2004[118]

2.Initiatives Taken By Working Group:

- National level study for e-waste assessment and approval of tor for national level study.
- Assessment of present scenario protections for future generation of PCs, mobile phone, TVs, ACs, refrigerators and other electronic equipment
- Prioritisation of cities for rapid assessments.

3.Base - Line And Data Management

- Work awarded to IRG South Asia Pvt Ltd. For development of e-waste manual has been completed and draft report has been submitted.
- Guidelines for assessment: approach and methodology and guidelines for training and capacity building.

5.4 WEEE (Waste Electrical and Electronic Equipment) Management Challenges in India

Widmer[119] identified difficulties specific to developing and industrializing countries in WEEE management after assessing management issues from China, India, and South Africa. These difficulties are summarized below:

- "although the quantity of indigenous e-waste per capita is still relatively small (estimated to be less than 1 kg e-waste per capita per year), populous countries such as China and India are already huge producers of e-waste in absolute terms;[120]
- these countries also display the fastest growing market for EEE;[121]
- some developing and transition countries are importing considerable quantities of e-waste. Some of them arrive as donations to help 'the poor' while others are mislabelled."

India, as one of the various developing countries, is witnessing an explosive consumption of electronic goods and equipment, and is on its way to becoming the choice destination for manufacturing.[122] The illegal import of e-waste from abroad added another 50 tonnes to the Indian-originated e-waste. Other findings of inventory study of MAIT and GTZ presented in December 2007 were that out of the materal entering the waste stream, only 5% is recycled by authorised recyclers. The rest is treated by the informal sector. But as only 6% of the organisations observed for the study had a disposal policy for e-waste, no one wonders why the material is entering the uncontrolled e-waste recycling stream. Considering the big potential of harming the environment and the human health, this study shows alarming results.[123] The challenges of managing e-waste in India are very different from those in other countries, developed or developing. When the density of e-waste in the country is on the rise, at present, there is no e-waste recycling plant in India.[124] While there can be several shared lessons, the complexity of the e-waste issue in India, given its vast geographical and cultural diversity and economic disparities, makes WEEE management challenges in India quite unique. A few of these are:

- Rapidly increasing e-waste volumes, both domestically generated as well as through imports. Imports are often disguised as second-hand computer donations towards bridging the digital divide or simply as metal scrap.[125]
- No accurate estimates of the quantity of e-waste generated and recycled.
- Low level of awareness amongst manufacturers and consumers of the hazards of incorrect e-waste disposal.
- Widespread e-waste recycling in the informal sector using rudimentary techniques such as acid leaching and open air burning resulting in severe environmental damage.[126]
- E-waste workers have little or no knowledge of toxins in e-waste and are exposed to serious health hazards.
- Inefficient recycling processes result in substantial losses of material value
- 'Cherry-picking' by recyclers who recover precious metals and improperly dispose of the rest.[127]

Table 3: Estimated Uses of Some Items in India

Items	Discard/Replace Rate
Mobile Phone	1-3 years
PC	Every 2 years
Camera	3-5 years
Television	10-15 years
Refrigerator	10-15 years
Washing Machine	10-15 years
IT Equipment	Very Fast (128)

Management of E-Waste

The electronic junks lie unattended in houses, offices, warehouses etc. and normally mixed with household wastes, which are finally disposed off at landfills. Therefore there is necessity for implementing management measures. In industries management of e-waste should begin at the point of generation. This can be done by waste minimization techniques, by sustainable product design and several other ways. [130]

Production-Process Modification

Changes can be made in the production process, which will reduce waste generation. This reduction can be accomplished by changing the materials used to make the product or by the more efficient use of input materials in the production process or both. Use of biodegradable/recyclable items will reduce e-waste. Most plastic waste is not biodegradable and will never decompose.[131] Recently companies are developing new plastics known as "bioplastics," which are biodegradable, reusable and compostable.[132] Improvements in the operation and maintenance of process equipment can result in significant waste reduction.

Hazardous materials used in either a product formulation or a production process may be replaced with a less hazardous or non-hazardous material.[133] For example, a circuit board manufacturer can replace solvent-based product with water-based flux and simultaneously replace solvent vapour degreaser with detergent parts washer.

Installing more efficient process equipment or modifying existing equipment to take advantage of better production techniques can significantly reduce waste generation. New or updated equipment can use process materials more efficiently producing less waste. Modifying existing process equipment can be a very cost-effective method of reducing waste generation.[134]

Modified Recycling Process

Recycling and the proper disposal of electronics is a major part of the solution to our growing e-waste problem.[135] Institutional infrastructures, including e-waste collection, transportation, treatment, storage, recovery and disposal, need to be established, at national and/or regional levels for the environmentally sound management of e-wastes.[136] These facilities should be approved by the regulatory authorities and if required provided with appropriate incentives. Establishment of e-waste collection, exchange and recycling centres should be encouraged in partnership with governments, NGOs and manufacturers.[137] Environmentally sound recycling of e-waste requires sophisticated technology and processes, which are not only very expensive, but also need specific skills and training for the operation. Proper recycling of complex materials requires the expertise to recognize or determine the presence of hazardous or potentially hazardous constituents as well as desirable constituents (i.e.those with recoverable value), and then be able to apply the company's capabilities and process systems to properly recycle both of these streams.[138]

Preventing e-waste from being landfilled mitigates the risk of these substances being released into the environment through landfill incineration or leachate leakage. Proliferation of hazardous materials could also be addressed at the recycler level.[139]

Although recycling can be a good way to reuse the raw materials in a product, the hazardous chemicals in e-waste mean that electronics can harm workers in the recycling yards, as well as their neighbouring communities and environment. In developed countries, electronics recycling takes place in purpose-built recycling plants under controlled conditions.[140]

In developing countries however, there are no such controls. Recycling is done by hand in scrap yards, often by children. There is a need to set up recycling plants preventing the environment and health hazards caused due to e-waste is a matter of serious concern.

Banning of Imports

E-waste is routinely exported by developed countries to developing ones, often in violation of the international law. Inspections of 18 European seaports in 2005 found as much as 47 percent of waste destined for export, including e-waste, was illegal. Mainland China tried to prevent this trade by banning the import of e-waste in 2000. The laws to protect workers and the environment are inadequate or not enforced. It is also cheaper to 'recycle' waste in developing countries. Demand in Asia for electronic waste began to grow when scrap yards found they could extract valuable substances such as copper, iron, silicon, nickel and gold, during the recycling process.[141] Not only has the United States government not ratified or complied with the Basel Convention, it has created policies that have actually encouraged the exportation of waste.[142] So there should be stringent laws in India to ban import of e-waste from the developed countries.

Reuse

It is a good way to increase a product's lifespan. Many old products are exported to developing countries. Although the benefits of reusing electronics in this way are clear, the practice is causing serious problems because the old products are dumped after a short period of use in areas that are unlikely to have hazardous waste facilities.[143] Reuse, in addition to being an environmentally preferable alternative, also benefits society. By donating used electronics, schools, non-profit organizations, and lower-income families can afford to use equipment that they otherwise could not afford.[144]

6.5 Capacity Building, Training and Awareness Programmes

An employee-training program is a key element of any waste reduction program. Training should include correct operating and handling procedures, proper equipment use, recommended maintenance and inspection schedules, correct process control specifications and proper management of waste materials.[145]

Lack of civic sense and awareness among city residents will be a major hurdle to keep e- waste out of municipal waste stream. Awareness raising programmes and activities on issues related to the environmentally sound management (ESM), health and safety aspects of e-wastes in order to encourage better management practices should be implemented for different target groups.[146] Collaborative campaigns are required to sensitise the users and consumers should pay for recycling of electronic goods. Consumers are to be informed of their role in the system through a labelling requirement for items. Consumers to be educated to buy only necessary products that utilize some of the emerging technologies (i.e. lead-free, halogen-free, recycled plastics and from manufacturers or retailers that will 'take-back' their product) to be identified through eco-labelling.[147]

E-Waste Policy and Regulation

It is the need of the hour to have an "e waste-policy" and national regulatory frame work for promotion of such activities. The main goal of any comprehensive e-waste policy should be a reduction, if not complete elimination, of the proliferation of hazardous materials into the environment through the e-waste stream.[148] The Policy shall address all issues ranging from production and trade to final disposal, including technology transfers for the recycling of electronic waste. Clear regulatory instruments, adequate to control both legal and illegal exports and imports of e-wastes and ensuring their environmentally sound management should be in place. There is also a need to address the loopholes in the prevailing legal frame work to ensure that e - wastes from developed countries are not reaching the country for disposal. The Port and the Custom authorities need to monitor these aspects. The regulations should prohibit the disposal of e-wastes in municipal landfills and encourage owners and generators of e-wastes to properly recycle the wastes. Manufactures of products must be made financially, physically and legally responsible for their products (Extended Producer Responsibility).[149]

Management Options in India

Considering the severity of the problem, it is imperative that certain management options be adopted to handle the bulk e-wastes. Designers should ensure that the product is built for re-use, repair and/or upgradeability. Stress should be laid on use of less toxic, easily recoverable and recyclable materials which can be taken back for refurbishment, remanufacturing, disassembly and reuse.[150] Recovery of metals, plastic, glass and other materials reduces the magnitude of e-waste. These options have a potential to conserve the energy and keep the environment free of toxic material that would otherwise have been released.[151] It is high time the manufactures, consumers, regulators, municipal authorities, state governments, and policy makers take up the matter seriously Following are some of the management options suggested for the government, industries and the public. the key role of manufactures and bulk consumers to shape and develop community participation

Responsibility and Role of Industries

All the actors along the product chain share responsibility for the lifecycle environmental impacts of the whole product system. These actors are the consumers, the suppliers, and the product manufacturers. ***Consumers*** can affect the environmental impacts of products in a number of ways: via purchase choices (choosing environmentally friendly products), via maintenance and the environmentally conscious operation of products, and via careful disposal (e.g., separated disposal of appliances for recycling).[152] ***Suppliers*** may have a significant influence by providing manufacturers with environmentally friendly materials and components. ***Manufacturers*** can reduce the life-cycle environmental impacts of their products through their influence on product design, material choices, manufacturing processes, product delivery, and product system support.[153]

The system design needs to be such that there are checks and balances, especially to prevent free riders. The goals of the product designer could include reducing toxicity and reducing energy use, streamlining product weight and materials, identifying opportunities for easier reuse, and more.[154] Manufacturers have to improve the design by: (i) the substitution of hazardous substances such as lead, mercury, cadmium, hexavalent chromium and certain brominated flame retardants;(ii) measures to facilitate identification and re-use of components and materials, particularly plastics; and (iii) measures to promote the use of recycled plastics in new products.[155]

Manufacturers should give incentives to their customers for product return through a "buy back approach" whereby old electronic goods are collected and a discount could be given on new products purchased by the consumer. All vendors of electronic devices shall provide take-back and management services for their products at the end of life of those products. The old electronic product should then be sent back to be carefully dismantled for its parts to be either recycled or re-used, either in a separate recycling division at the manufacturing unit or in a common facility.[156]

Responsibilities of the Government

Governments should also undertake certain responsibilities. It should set up regulatory agencies in each district, which are vested with the responsibility of co-ordinating and consolidating the regulatory functions of the various government authorities regarding hazardous substances.

1. Governments should be responsible for providing an adequate system of laws, controls and administrative procedures for hazardous waste management (Third World Network. 1991). A comprehensive law that provides e-waste regulation and management and proper disposal of hazardous wastes is required. Such a law should empower the agency to control, supervise and regulate the relevant activities of government departments.[157]

2. Governments must encourage research into the development and standard of hazardous waste

management, environmental monitoring and the regulation of hazardous waste-disposal.[158]

3. Governments should enforce strict regulations against dumping e-waste in the country by outsiders. Where the laws are flouted, stringent penalties must be imposed. In particular, custodial sentences should be preferred to paltry fines, which these outsiders / foreign nationals can pay.
4. It is the need of the hour to have an "e waste-policy" and national regulatory frame work for promotion of such activities.[159] Governments should encourage and support NGOs and other organizations to involve actively in solving the nation's e-waste problems.[160]
5. Uncontrolled dumping is an unsatisfactory method for disposal of hazardous waste and should be phased out.
6. Governments should explore opportunities to partner with manufacturers and retailers to provide recycling services.
7. The government could also intervene to discourage the use of sweatshop labour in the processing of e-waste.[161]

Responsibilities of the Citizen

Waste prevention is more preferred to any other waste management option including recycling. The citizens should undertake certain initiatives for management of e-wastes. Donating electronics for reuse extends the lives of valuable products and keeps them out of the waste management system for a longer time. But care should be taken while donating such items i.e. the items should be in working condition.

E-wastes should never be disposed with garbage and other household wastes. Consumers wishing to dispose of products containing hazardous materials would be required to drop them off with qualified collection centers which would properly process the hazardous materials.[162] Customers should opt for upgrading their computers or other electronic items to the

latest versions rather than buying new equipments. While buying products they should take into account its certification by regulatory authorities or products having offer leasing or take back options, energy efficient etc.[163]

Conclusion

Solid waste management, which is already a mammoth task in India, is becoming more difficult by the attack of e-waste. There exists a vital need for a detailed evaluation of the current and future scenario including quantification, characteristics, existing disposal practices, environmental impacts etc. The systems in place both here and abroad are inconsistent because they focus too much of the burden of handling e-waste on too few parties. Institutional infrastructures, including e-waste collection, transportation, treatment, storage, recovery and disposal, need to be established, at national and/or regional levels for the environmentally sound management of e-wastes. Establishment of e-waste collection, exchange and recycling centers should be encouraged in partnership with private entrepreneurs and manufacturers.

There is a need for policy level interventions which should include development of e-waste regulation, control of import and export of e-wastes and facilitation in development of infrastructure. Effective programs providing incentives for producers to design products that are less wasteful, contain fewer toxic components, and are easier to disassemble, reuse, and recycle can help a lot in reducing the wastes. It should set targets for collection and reuse/recycling, impose reporting requirements and include enforcement mechanisms and deposit/refund schemes to encourage consumers to return electronic devices for collection and reuse/recycling. End-of life management should be made a priority in the design of new electronic products. The amount and toxicity of e-waste is growing to epic proportions and if not checked, will cause irreversible harm. So, there is a need for the creation of better management techniques and their effective implementation which in turn will generate public awareness. Only when this is in place, will the detrimental effects of e-waste begin to be curbed.

End Notes

1. See Computer Take Back Campaign, Poison PCs and Toxic TVs 8 (2004), available at http://www. computertakeback.com/ docUploads/ppcttv 2004%2 Epdf?CFID= 17553870& CFTOKEN= 11135182 accessed on 4/03/2010
2. Courtney Rob, *'Evolving Hazardous Waste Policy for the Digital Era'*, 25 Stanford Environmental Law Journal. 199, 227 (2006).
3. Billinghurst M. Betsy, *'E-Waste: A Comparative Analysis of Current and Contemplated Management Efforts by the European Union and the United States'*, 16, Colorado Journal of International Environmental Law & Policy 399, 400 (2005).
4. Ezroj Aaron, *'How The European Union's Weee & Rohs Directives Can Help The United States Develop A Successful National E-Waste Strategy'*, 28 Virginia Environmental Law Journal,2006,45
5. *Ibid.*
6. Kutz Jennifer, *'You;ve Got Waste: The Exponentially Escalating Problem of Hazardous e-waste'*, Villanova Environmental Law Journal, 17 ,307,2006.
7. Available at http://www.calrecycle.ca.gov/Electronics/Whatis Ewaste accessed on 12/02/2010
8. Available at www.eWasteGuide.com/WEEEBasics accessed on 13/ 02/2010
9. *Ibid.*
10. Antrekowitsch, H., Potesser, M., Spruzina, W. & Prior, F. (2006), *'Metallurgical recycling of electronic scrap. In: Howard, S.M.,Antrekowisch, H. Potesser, M. Spruzina, W and Prior, F,* EPD(Environment Production Declaration) Congress 2006, pp. 889-908. The Minerals, Metals and Materials Society, TMS, Warrendale, Penninslavia, USA.
11. *Ibid.*
12. *supra* note 8 (reffering to end note 8)

13. Available at http://india.ewasteguide.info/Initial accessed on 14/02/2010
14. *supra* note 8
15. Available at http://www.iimm.org/knowledge_bank/9_e-weste-management.htm accessed on 2/03/2010
16. Dickenson, J. (2006), *'Electronic Signals: a Year into the EU WEEE Directive'*. Waste Management World. July–August, 2006, pp. 37–47.
17. Hageluken, C. (2006), Recycling of electronic scrap at Umicore's precious metals refining. *Acta Metallugica Slovaca,* **12**, 111–120.
18. Fichter, K. (2003), *'E-commerce: sorting out the environmental consequences',* Journal of Industrial Ecology, **6**, 25–41.
19. AIIA & Planet Ark Consulting (2005) AIIA: *'E-waste Program Development Phase'*. Report for discussion and feedback. AIIA and Planet Ark Consulting. June, 2005.
20. *supra* note 14
21. Available at www.efymag.com accessed on 15/02/2010
22. Available at http://www.jimandaz.com/1/e_waste_ hazardous_heatlh.htm accessed on 2/03/2010
23. *Ibid.*
24. Available at http://wmr.sagepub.com/cgi/content/refs/25/6/489 accessed on 25/02/2010
25. *Ibid.*
26. Available at http://en.wikipedia.org/wiki/Electronic_ waste# Definition accessed on 12/02/2010
27. Agarwal R. (1998) *'India: The World's Final Dumpyard!'*, January, Basel Action News, Vol.1 available at www.ban.org accessed on 14/02/2010
28. *supra* note 18
29. Shobha B.S.Devi, R.K. Kamble S. V (2004), *'E-Waste: The Hidden harm of Technological Revolution'*, Journal IAEM (International Association of Emergency Managers), Vol.31, p.196-205.
30. *supra* note 21
31. Available at http://ewasteguide.info/node/219 accessed on 25/02/2010
32. *Ibid.*

33. *Ibid.*

34. Available at http://www.earthtimes.org/articles/show/310528,toxic-electronic-waste-poses-increasing-risk-for-developing-world.html accessed on 2/03/2010

35. BAN/SVTC (2002) *'Exporting Harm: the High Tech Trashing of Asia'*, The Basel Action Network and Silicon Valley Toxics Coalition. February 25, 2002

36. Available at http://ezinearticles.com/?Electronic-Waste—And-it-Was-Really-As-Dangerous-As-This&id=3434748 accessed on 23/02/2010

37. Templeton J. Nicola, *'The Dark Side Of Recycling And Reusing Electronics: Is Washington's E- Cycle Program Adequate?'*, 7 Seattle Journal for Social Justice. 763,2009,Spring/Summer

38. *Ibid.*

39. Clapp Jennifer, Toxic Exports: The Transfer of Hazardous Wastes from Rich to Poor Countries 1 (2001) Cornell University Press,5th ed.(emphasis in the original).

40. Scharnhorst, W., Althaus, H.-J., Classes, M., Jolliet, O. & Hilty, L.M.(2005), *'The end of life treatment of second generation mobile phone networks: strategy to reduce the environmental impact'*, Environmental Impact Assessment Review, **25**, 540–566.

41. Craig Lorch of Seattle's Total Reclaim is an environmentalist and businessman who is opposed to the idea of exporting e-waste to developing countries and has developed a means of managing leaded CRT glass locally. However, he said, "Every day we [recyclers] must make a choice between spending money to disassemble and manage material here or simply load it into a shipping container and sell it offshore. Why would a good business person allow spending [three] to [four] dollars to disassemble and handle material domestically when the same material can be sold offshore for [three] bucks. It's a [six] dollar swing—that's a make or break difference for a recycler." ExportinG Harm, See Basel action network & Silicon Valley Toxics Coalition, Exporting Harm: The High-Tech Trashing of Asia 5 (2002), available at www.ban.org/E-waste/technotrashfinalcomp.pdf accessed on 3/03/2010; Gordon Fairclough, E-Waste' From Computers Discarded in West Turns Up in China's Exported Trinkets, Wall Street Journal, July 12, 2007, at Bl; Our E-waste Comes Back to Haunt Us, (American Public Media radio broadcast Nov. 14, 2007), available at http://marketplace.publicradio.Org/display/web/2007/11/14/consumed5_pm_1 accessed on 21/02/2010

42. *supra* note 2
43. *supra* note 41 Basel Action Network
44. Plepys, A. (2002) '*Implication of globalization and new product policies for the suppliers from developing countries*'. In: Proc. International Symposium on Electronics and the Environment 2002, IEEE, May 6–9, 2002, San Francisco, CA, USA, pp. 53–58.
45. Aside from the toxic elements e-waste contains, there are also valuable materials and precious metals that can be recovered and reused, for example glass, plastics, steel, aluminum, as well as gold, silver, platinum, and copper. In fact, one metric ton of computer scrap contains more gold than seventeen tons of gold ore. In 1998, the amount of gold recovered from electronic scrap in the United States was equivalent to that recovered from two million metric tons of gold ore. U.S. GEOLOGICAL SURVEY
46. *supra* note 2,3
47. U.S. Geological Survey, U.S. Dep't of the interior, Usgs Fact Sheet FS-060-01, Obsolete Computers, "Gold mine," OR High-Tech Trash? Resource Recovery From Recycling (2001), available at http://pubs.usgs.gov/fs/fs060-01/fs060-01.pdf accessed on 4/03/2010
48. *supra* note 2
49. Available at http://india.ewasteguide.info/e_waste_ generation_ in_india_gtz_mait_study_results accessed on 3/03/2010
50. *supra* note 21
51. Available at http://www.e-waste.in/weee_basics/weee_ hazards / visited on 25/02/10
52. Joseph Kurian, '*Electronic Waste Management In India–Issues And Strategies*', Sardinia 2007, Eleventh International Waste Management and Landfill Symposium S. Margherita di Pula, Cagliari, Italy; 1 - 5 October 2007
53. Silicon Valley Toxics Coalition, Poison PCs and Toxic TVs: California's Biggest Environmental Crisis That You've Never Heard Of 8 (2004) (noting the rapid increase in e-waste production), available at http:// svtc.etoxics.org/site/DocServer/ppc-ttv1.pdf?docID=124 accessed on 24/02/2010
54. U.S. Environmental Protection Agency, Frequent Questions, available at http://www.epa.gov/epaoswer/hazwaste/recycle/ ecycling/faq.htm#concern visited on 5/03/2010

55. Available at http://technology.findlaw.com/articles/00006/010271.html accessed on 17/02/2010

56. Technology Transfer Network Air Toxics Web Site, Lead Compounds, U.S. Envtl. Protection Agency (2000), http://www.epa.gov/ttn/uatw/hlthef/lead.html visited on 25/02/10.

57. Occupational Safety & Health Admin., U.S. Dep't of Labor, Cadmium Health Effects (2005), available at http://www.osha.gov/pls/oshaweb/owadisp.show_ document?p_table= PREAMBLES&p_id=819 (studying effects of occupational exposure to cadmium)visited on 2/03/10

58. Office of Research Facilities, National Institutes of Health, Mercury Health Hazards (2006), available at http:// orf.od.nih.gov/Environmental+Protection/Mercury+Free/Mercury Health Hazards. htm visited on 02/03/10

59. *supra* note 56

60. Schut H. Jan, *'Recycling E-Plastics: New Material Stream Brings Its Own Set of Problems, Plastics Tech.'*, Aug. 2007, available at http://www.ptonline.com/articles/200708fa2.html accessed on 4/03/2010. For a comprehensive study of the adverse effects of plastics, see EcologyCenter.org, Adverse Health Effects of Plastics, available at http://www.ecologycenter.org/factsheets/plastic health effects.html visited on 03/03/10

61. Michael Belliveau & Stephen Lester, Ctr. for Health, Env't, & Just. and Envtl. Health Strategy Ctr., PVC: Bad News Comes in 3's - The Poison Plastic, Health Hazards and the Looming Waste Crisis (2004), available at http://www.watoxics.org/files/pvc-bad-news.pdf. visited on 03/03/10

62. Iowa Dep't of National Research :Environmental and Health of Electronic Waste Hazards", available at http://www.iowadnr.com/waste/recycling/hazards.html#flame visited on 03/03/10

63. Agency for Toxic Substances & Disease Registry, U.S. Dep't of Health & Hum. Servs., ToxFAQs for Beryllium (2002), available at http:// www.atsdr.cdc.gov/tfacts4.pdf visited on 02/03/10

64. *supra* note 1

65. National Safety Council, Lead Poisoning (Dec. 23, 2004), available at http:// www.nsc.org/library/facts/lead.htm (stating effects lead exposure has on children under age six) visited on 02/03/10

66. *supra* note 57

67. See U.S. Environmental Protection Agency, Health Effects of Mercury, http://www.epa.gov/mercury/effects.htm visited on 02/03/10

68. Alliance for Safe Alternatives , The American People's Dioxin Report, available at http://www.safealternatives.org/report.html#Chapter%208 visited on 02/03/10 (studying human health effects of dioxins).

69. Environmental Justice Advocates, Dioxin, http:// www.ejnet.org/dioxin/ accessed on 4/03/2010(referencing EPA's Draft Dioxin Reassessment and research conducted by National Toxicology Program). A discussion of the research can be found at http://ntp-server.niehs.nih.gov/ accessed on 4/03/2010

70. Ramachandra T.V., Saira Varghese K, '*Environmentally Sound Options For E-Wastes Management;*', Energy and Wetlands Group, Center for Ecological Sciences, Indian Institute of Science, Bangalore. Published by : Envis Journal of Human Settlements, March 2004.

71. *supra* note 6

72. Press Release, United Nations University, UN, Industries, and Others Partner to Create World Standards for E-Scrap Recycling, Harvest of Valuable Components, U.N. Doc. MR/E11/07 (Mar. 6, 2007), available at http:// www.unu.edu/media/archives/2007/files/mre11-07.pdf. visited on 03/03/10

73. Thakker Nisha, '*India's Toxic Landfills: A Dumping Ground for the World's Electronic Waste*', 6 Sustainable Dev. L. & Pol'y 58, 58 (2006).

74. *supra* note 1

75. Cobbing Madeline, '*Greenpeace International, Toxic Tech: Not In Our Backyard - Uncovering The Hidden Flows Of E-Waste*', 15 (2008) At 48

76. CII (2006). "E-waste management", Green Business Opportunities, Vol.12, Issue 1,Confederation of Indian Industry, Delhi.

77. Jim Puckett & Ted Smith., Exporting Harm: the High-Tech Trashing of Asia, 1, 2 (Jim Puckett & Ted Smith eds., 2002) [hereinafter Puckett, Exporting Harm], available at http://svtc.etoxics.org/site/DocServer/technotrash.pdf? docID=123 accessed on 3/03/2010 at 26

78. Catherine K. Lin Jennifer L. Fordyce., '*Globalization, Extended Producer Responsibility and the Problem of Discarded Computers in China: An Exploratory Proposal for Environmental Protection*', 14 Geo. Int'l Envtl. L. Rev. 525, 526 (2002).at 553

79. *supra* note 70
80. Widmer R, Heidi Oswald-Krapf , Deepali Sinha-Khetriwal, Max Schnellmann, Heinz Bo¨ni (2005), '*Global perspectives on e-waste*', Environmental Impact Assessment Review 25 436–458
81. Available at http://www.hindu.com/2007/02/19/stories/2007021901050900.htm accessed on 2/03/2010
82. High Tech Business Begets High Tech Environmental Problems - Buchanan Ingersoll, P.C.; Antoineette R Stone; January 2000
83. The Basel Convention and Transboundary Movements of Hazardous Wastes - Jonathan Krueger, Energy and Environmental Programme, Briefing no: 45 May 1998, The Royal Institute of International Affairs
84. The Basel Ban - The Pride of the Basel Convention - An update on Implementation and Amendment - September 1995 available at www.greenpeace.org accessed on 4/03/2010
85. Available at http://www.e-waste.in/weee_policy/international_legislation/ accessed on 3/03/2010
86. Available at http://www.tematelecom.net/upload_images/296E-waste-DILIP.pdf accessed on 2/02/2010
87. European Parliament and Council Directive 2002/95/EC, 2003 O.J. (L 37) 19 [hereinafter RoHS Directive] (on the Restriction of the Use of Certain Hazardous Substances in Electrical and Electronic Equipment); European Parliament and of the Council Directive 2002/96/EC, 2003 O.J. (L 37) 24
88. *Ibid.*
89. Available at http://ewasteguide.info/swiss-e-waste-programme accessed on 3/03/2010
90. Available at http://www.e-waste.ch/en/policy-and-legislation/index.html accessed on 3/03/2010
91. *supra* note 21
92. Radha G. (2002). '*A Study of the Performance of the Indian IT Sector*' at www.nautilus.org accessed on 3/03/2010
93. Mehra H.C. (2004). PC waste leaves toxic taste, The Tribune, 22nd March.
94. *Ibid.*
95. supra note 13

96. Shobana Ramesh and Kurian Joseph (2006). *'Electronic waste generation and management in an Indian city'*, Journal of Indian Association for Environmental Management, Vol. 33, No.2, pp 100-105

97. *supra* note 15

98. *supra* note 49

99. Sergio J. and Tohru M. (2005) Waste management of electric and electronic equipment: comparative analysis of end-of-life strategies, J. Mater. Cycles Waste Manag. 7:24–32, *supra* note 96, 80

100. *supra* note 27

101. Annual Report 2003, (Manufacturing Association of Information Technology)MAIT

102. Available at http://india.ewasteguide.info/e_waste_actors accessed on 24/02/2010

103. Toxics Link report: Scrapping the Hi-tech Myth: Computer Waste in India, Feb. 2003: www.toxicslink.org accessed on 23/02/2010

104. *supra* note 102

105. *Ibid.*

106. *Ibid.*

107. *Ibid.*

108. *Ibid.*

109. *supra* note 29

110. Packaging, Recycling and Solid Waste – Executive Summary, Policy Study No 223, June 1997 of Reason Public Policy Institute, Los Angeles

111. *supra* note 76

112. Avaialble at www.cpcb.nic.in accessed on 3/03/2010

113. *Ibid.*

114. DIT (2003). Environmental management for Information Technology industry in India, Department of Information Technology, Government of India, pp.122-124.

115. *Ibid.*

116. Available at http://www.ewa.co.in/ accessed on 23/02/2010

117. Available at http://www.swlf.ait.ac.th/UpdData/International/NRIs/Electronic%20waste%20management%20in%20India.pdf accessed on 4/03/2010
118. Available at http://www.ciol.com/EDA/Enterprise/Feature/E-waste-management-The-call-for-a-better-tomorrow/22408105382/0/ accessed on 26/02/2010
119. *supra* note 80
120. Toxic Links (2003) *Scrapping the High-tech Myth: Computer Waste in India.* Toxic Links, Delhi, India.
121. Ravi, V., Shanker, R. & Tiwari, M. K. (2005). *'Analyzing alternatives in reverse logistics for end-of-life computers: ANP and balanced scorecard approach'*, Computer and Industrial Engineering, **48**: 327–356.
122. *supra* note 81
123. *supra* note 49
124. Available at http://www.expressindia.com/news/fullstory.php?newsid=55274 accessed on 4/03/2010
125. *supra* note 76
126. *supra* note 93
127. Available at http://www.e-waste.in/weee_initiatives/ accessed on 23/02/2010
128. *supra* note 21
129. *Ibid.*
130. Available at http://india.ewasteguide.info/Strategy_ewaste visited on 02/03/10
131. David Aftandilian, The Age of Green Plastics May Be Coming Soon, Conscious Choice (May 2002), http:// www.consciouschoice. com/2002/cc1505/note1505.html visited on 05/03/10
132. *Ibid.*
133. European Parliament and Council Directive 2002/95/EC, 2003 O.J. (L 37) 19 [hereinafter RoHS Directive] (on the Restriction of the Use of Certain Hazardous Substances in Electrical and Electronic Equipment); European Parliament and of the Council Directive 2002/96/EC, 2003 O.J. (L 37) 24
134. *supra* note 70
135. *supra* note 55

136. *supra* note 76

137. *Ibid..*

138. *Ibid.*

139. Avaialble at http:// www.epa.gov/dfe/pubs/about/dfe-highlightso0b.pdf visited on 05/03/10

140. Available at http://www.greenpeace.org/international/campaigns/toxics/electronics/where-does-e-waste-end-up visited on 03/03/10

141. *Ibid.*

142. *supra* note 77

143. *supra* note 140

144. *supra* note 70

145. *Ibid.*

146. Asia-Pacific Regional Scoping Workshop on Environmentally Sound Management of Electronic Wastes, 19-22 November, 2002, Tianjin, China The Conference was organized by Basel Convention Regional Centre in Beijing and Secretariat of Basel Convention.

147. *Ibid.*

148. *supra* note 133

149. *supra* note 96

150. *supra* note 70

151. *Ibid.*

152. Dana Joel Gattuso, Mandated Recycling of Electronics: A Lose-Lose-Lose Proposition, Competitive Enterprise Institute: Issue Analysis, Feb. 1, 2005, available at http://www.cei.org/pdf/4386.pdf

153. *supra* note 70

154. *supra* note 96

155. Green Product Design, GreenBiz.com, available at http://www.resourcesaver.org/file/toolmanager/O16F7588.pdf visited on 05/03/10

156. *supra* note 99

157. Third World Network. 1991. Toxic Terror: Dumping of Hazardous Wastes in the Third World, Third World Network, Malaysia.

158. *supra* note 70

159. Ammons J and Sarah B. (2003) *'Eliminating E-waste: Recycling through Reverse Production'* at www.lionhrtpub.com accessed on 02/03/10

160. Freeman M. H. 1989. Standard Handbook of Hazardous Waste Treatment and Disposal, McGraw-Hill Company, USA.

161. Available at http://ban.org/country_status/country_ status_chart.html accessed on 3/03/2010

162. *supra* note 152

163. *supra* note 70

Chapter 10

Status of E-Waste Management in Delhi

Sirajuddin Ahmed, Mohammed Shakeel and Rashmi Makkar Punwar

Introduction

The term electronics encompasses a wide range of home and business electronic goods, including televisions, computer monitors, computer peripherals, audio and stereo tapes, VCR's, DVD players, video cameras, telephones, fax and copy machines, cellular phones, wireless devices etc. Discarded electronics which is referred to as end of life of electronic products, electronic wastes, E-waste or waste electrical and electronic equipment. (WEEE)-are generated, when users or owners of the product decide that they no longer want them .E-waste is popular informal name of electronic products which are nearing or have reached the end of useful life [8]. Recent policy changes in India have lead to an influx of leading multinational companies to setup electronic manufacturing facilities and R & D centers for hardware and software. This has undoubtly helped the Indian economy to grow faster. This on the other hand has given rise to increase in the consumption rate of electronic items especially computers which ultimately add up to the waste load. The newer models are available at affordable prices, which causes the net higher rate of obsolescence leading to growing piles of E-waste [10]. The necessity of E-waste management arises due to complexity and toxicity of compounds present in computers. E-waste contains over 1000 different substances and chemicals, many of which are toxic and are likely to create serious problems for the environment and human health if not handled properly [7].

Necessity Of E-Waste Management

Quantity of E-waste

According to report submitted by United Nations Environment Programme (UNEP) in February 2010, currently (base year 2007) E-waste generation in India is over 100,000 tonnes from refrigerators 275,000 tonnes from T.V's, 56,3000 tonnes from personal computers, 4700 tonnes from printers, and 1700 tonnes from mobile phones. It is predicted that by 2020 E-waste from old computers will have jumped by 500% in India from the level in 2007. The E-waste from mobile phones in the same period will be 18 times higher and from refrigerators it would be 2-3 times [15]. This is not the only nerving problem but in addition to this large scale unethical export of E-waste by industrialized nations to developing countries like India, China, and Pakistan is shifting the onus of development to communities that are ill equipped to deal with such wastes. According to a study compiled by the Graduate School of Industrial Administration of Carnegie Mellon University, in year 2002 estimated that in year 2002, 12.75 million computer units went to recyclers in U.S. Based on this it was estimated that e-waste is moving at the rate of 80% to Asian countries. The total amount sent to Asia was about 10.2 million units[9].

Complexity of E-waste

E-waste comprises of a large number of electronic products. Manufacturing of which involves a number of hazardous substances and chemicals like heavy metals which eventually become E-waste after completing their life cycle.

E-waste contains a witches brew of toxins such as lead and cadmium in circuit board, Lead oxide in Cadmium Cathode Ray Tubes (CRT'S), Mercury in switches and flat screen monitors, Cadmium in computer batteries, Polychlorinated Biphenyls (PCB'S) in older capacitors and transformers and Brominated Flame retardants in Printed Circuit Boards, plastic casings, cables and Polyvinylchloride (PVC) cable insulation materials that release highly toxic dioxins and furans when burnt. [2] Land filling of E-wastes can lead to the leaching of lead into ground water. If the CRT is crushed and burned, it emits toxic fumes into the air [7].

Electronic waste contains several rechargeable battery types, all of which contains toxic substances that can contaminate the environment when burned in incinerators or disposed off in landfills. The cadmium from one mobile phone battery is enough to pollute 600 cubic meter of water [1]. The quantity of cadmium in landfill sites is significant and considerable toxic contamination is caused by the inevitable medium and long term effects of cadmium leaking into surroundings soil [12]. Plastics are highly inflammable and a major part of computer is made up of plastics, which release harmful dioxins, furans, polychlorinated biphenyls and brominated flame retardants.

Impacts of informal recycling in Delhi

Computer waste is often routed through main stream channels such as custom ports, computer service centers etc. and pass through grey market. According to an industry lobby it is estimated that over 2000 trucks dump around 12,000 tonnes of E-waste in the city per day, at different places in Delhi: Turkman Gate, Shastri Park, Loni, Seelampur and Mandoli [5].

Delhi employs 25000 workers approximately at its scrape yards, which handle the E-waste unaware of harmful effects of the toxins released from them. All markets of E-waste and recycling units for computer waste are situated in densely populated areas of Delhi. The computers are recycled in backyards by unsophisticated, rudimentary and dangerous methods practiced by untrained, unprotected and unaware people. Computers are manually dismantled by using a hammer and a chisel. Each component is separated from its unit. Monitors are much sought after by scrap dealers as they contain good amount of copper besides circuit boards and picture tubes. Wires are scraped and stripped off manually to separate copper. Integrated circuits are removed from circuit boards manually. *Katiyas* are experienced to find out gold and silver used in soldering operations which they remove by cutting those parts or by heating on stoves

Printed wiring boards are treated for hours with sulphuric acid to remove copper embedded in it by precipitating it in complex form. Waste originates from acids slag, ashes and glass obtained from CRT smashing during the recovery process. Waste components which do not have any resale or reuse value are openly burnt or disposed off in open dumps. The economics of backyard recycling units themselves is very complex. Each component or recovered item has a market value. Most of the E-waste recycling is done by small enterprises which are working illegally and informally in this field. Generally the scrap dealers have extended their work to computer dismantling. Poor workers are working on bare minimum wages and in close vicinity of extremely dangerous chemical emissions and physical conditions. They are generally operating from border areas of Delhi where this business can not be intervened by Delhi Government *authorities.*

Current status of E- waste management in Delhi

A wide range of environmental policies and legislation provide guide lines for import and export and handling of computer waste. But there are certain loopholes in the current legal system which has given rise to the mountains of junk in Delhi. A number of steps and actions have been taken by Government to provide sound management of electronic waste.Some of them are:

1. A National workshop was conducted by MoEF in New Delhi,December 10, 2008.

2. Rapid assessment of E- waste in Delhi and identification of E-waste handling hotspots have been done [6];

3. The hazardous waste management and handling rules as proposed in 1998 was amended in 2008 for toxic content-registration mandatory for recyclers;

4. Guidelines notified by Central Pollution Control Board (CPCB) in April 2008,identify and recognize; a)Producer's responsibility, b) Restriction on hazardous substances (RoHS), c) Best practices d)Insight into technologies for various levels of recycling.[4]

5. The country has 12 authorized recyclers including E-Parisara, Ash, Eco-reco, Ramky group, and Attero. Attero has opened its E-waste facility in Roorkee Uttarakhand in February 2010, where the E-waste of Delhi's dump yards will also be treated [3];

6. To manage E-waste, MCD and NDMC have been directed to have specially designed E-waste bins and to install them at accessible places to residents. NDMC has a system of floating tenders to auction discarded electronic items for safe disposal and recycling [*14*];

7. Various plans and programs are executed by the NGO,s for collection of E- waste from small house holds E- waste bins were installed at St, Paul school with the help of students and the collected -waste was sent to authorized dealers [*16*];

8. Actions are taken by CPCB for rapid assessment of E- waste generated and handled in Delhi. This will provide sufficient data base for future plans of handling it;

9. A structured proposal on E -waste legislation has been advocated by Vinnie Mehta (Executive director of MAIT {Manufacturers Association and Information Technology}) and is under active consideration by the members of parliament;

Although awareness and readiness for implementing improvements is increasing rapidly, the major obstacles to manage -waste safely and effectively remains the need of the hour.

Waste Management Orientation Measure

Waste Management orientation is also an important issue and following measures should be adopted to ease out the alarming situation:

1. *Ban hazardous waste imports:*

There is no reason for the poor of the world to bear the burden of environmental risk, particularly when they have not used it or taken benefits of it.

2. *Make producer responsible:*

The producer must be responsible on the principle of extended producer's responsibility (EPR) over the entire life cycle of the product. This will also enforce the producer to makeshift the technology to design for disassembly and design for recyclability.

Computer manufacturing involves computer aided design (CAD) tools in designing process, so data can incorporated into CAD software for environmental considerations. This will also enhance the possible material substitution for the safe disposal [*13*].

3 *Consumer oriented measures:*

(a) The product lease approach should be incorporated in which the product sold remains the property of producer. It is therefore producer's interest to increase the life span of product and develop efficient reuse and recycling strategies.

(b) Changing consumption patterns by choosing eco-efficient, long life products and stop buying useless products.Consumer must be informed and should be responsible about the hazardous material present and the consequences of unsafe disposal through unauthorized handlers.

Conclusion

Developed countries have been practicing fully automated recycling technology, which involves a high cost whereas in India and especially in Delhi E-waste recovery and recycling is mainly done in unorganized manner and mostly manually by untrained workers. As the possibility of 90% of resource recovery is claimed by E-parisara Bangalore [*11*], and there is a bright possibility of recovering the maximum from electronic junk. However it is not possible without the support and efforts of consumer who should bear the primary responsibility of making E- waste reach the organized handlers. E- waste is actually not a waste it is E- wealth from secondary resource for metals and non- metals. Resource recovery can be carried out without harming ecology and environment if formal recycling units with sound technology come up in Delhi also. If a judicial combination of trained manual labor and technical support is involved in handling the E- waste in Delhi our heaps of electronic junk can be the source of resources of wealth generation.

References

1. Trick (2002), *"A Mobile is not Just for Christmas"*; 24th December 2002, http://newsbbc.co.ok accessed on August 2005.
2. Parvinder Khanuja, Saroj Mahajan, Shakeen Mishra (2005), *"A New Man MadeSsource of Pollution –Electronic Wastes"* ,Department of Chemistry and Botany, S.N.P.G. College Khandwa; proceedings of ICCE-2005(Internatioanal Conference on Computers in Education).
3. ITVN network (2010), *"Attero Inaugurated Facility to Extract Metals from e- waste"*; Jan23, 2010.
4. *Central Pollution Control Board " Guidelines"* www.cpcb.nic.in
5. *Delhi an E-waste Dumping Yard"* August 18th 2008, www. thailandian.com

6. Kurian Joseph(2006) *"Electronic Waste Management in India -Issues and Strategies"*, Journal of Indian Association of Management, Vol 33.

7. Environmental Engineering, University of the Philippines, *"Electronic waste: A threat in the future"* Environmental Engineering, University of the Philippines, Diliman's; www. inter discipliplimary.net

8. Ramachandra T.V, Saira Varghesek *(2004), "Environmentally sound options of E-waste management"*, Envis Journal of Human Settlements,March 2004

9. *Toxics link "E-waste in India*: Report", toxic links,www.toxic links.org

10. AIR CMDE (RETD) P.D.Badoni , VSM *"E-waste management in India"*, Jan 2009, Electronic for you,www.efymag.com.

11. Parthasarathy P,Keshav A Bulbule and Anantha Murthy K.S.(2008), *"E-waste recycling -Best option for resource recovery and sustainable environment*; Research journal of chemistry and environment; vol.12(1) March (2008)

12. *"Mobile phones recycling at Envocare* (2001),www.envocare.co.uk, accessed on 28th august 2005

13. *"Recommendations on reductions of waste amounts and emissions from WEEE"*, ewasteguide.info

14. *NDMC (2010),"To manage E waste Govt. uses segregated disposal"*, 10 Feb, 2010,www.ndmc.gov.in

15. *UNEP,"Urgent needs to prepare developing countries for surge in E-waste"* ; Report by UNEP , www.unep org

16. *"Workshop on E-waste"*; students environment. blog spot .com; march 2008

Chapter 11

Solid Waste: Management and Related Health Hazards-- An Indian Perspective

Anubhuti Dungdung and Ekta Gurjar

Introduction

Since the beginning, humankind has been generating waste, be it the bones and other parts of animals they slaughter for their food or the wood they cut to make their carts. With the progress of civilization, the waste generated became of a more complex nature.[1] At the end of the 19th century not only did the air get more and more polluted but the earth itself became more polluted with the generation of non- biodegradable solid waste. The increase in population and urbanization was also largely responsible for the increase in solid waste.[2]

In living organisms, waste is the unwanted substances or toxins that are expelled from them. More commonly, waste refers to the materials that are disposed of in a system of waste management.[3]

According to the Basel Convention of the United Nations Environment Program, Waste are:

"Substances or objects which are disposed of or are intended to be disposed of or are required to be disposed of by the provisions of national law"[4]

As produced by the United Nations Statistics Division (UNSD) of the United Nations Environment Program:

"Wastes are materials that are not prime products (that is products produced for the market) for which the generator has no further use in terms of his/her own purposes of production, transformation or consumption, and of which he/she wants to dispose. Wastes may be generated during the extraction of raw materials, the processing of raw materials into intermediate and final products, the consumption of final products, and other human activities. Residuals recycled or reused at the place of generation are excluded."[5]

Pongrácz and Pohjola, scholars, have defined Waste as,

"Things with well-defined Purpose, and acceptable Performance, but their users fail to use them for the intended Purpose."[6]

The UK Environment Agency classifies waste as either controlled waste or non-controlled waste[7]. Controlled waste includes waste generated from households (solid waste), commercial and industrial organizations and from construction and demolition. Non-controlled waste includes waste generated from agriculture, mines and quarries and from dredging operations. Solid waste (SW) consists of many different things including food and garden waste, paper and cardboard, glass, metals, plastics and textiles.[8] Solid waste (SW), also called urban solid waste, is a waste type that includes predominantly household waste (domestic waste) with sometimes the addition of commercial wastes collected by a municipality within a given area. They are in either solid or semisolid form and generally exclude industrial hazardous wastes.[9]

Florida Public Law defines Solid Waste as:

"sludge unregulated under the Federal Clean Water Act or Clean Air Act; sludge from a waste treatment works, water supply treatment plant, or air pollution control facility; or garbage, rubbish, refuse, special waste, or other discarded material, including solid, liquid, semi-solid, or contained gaseous material resulting from domestic,

industrial, commercial, mining, agricultural, or governmental operations."[10]

From this definition, "solid waste" is comprised by law of materials that are a mixture of liquids, gases, and solids. In addition, a small amount of hazardous waste is allowed under the definition of solid waste.[11] This hazardous portion of solid waste is proposed as one of the primary reasons for the political opposition and attention that the solid waste industry faces today, as well as the sheer volume produced and how to dispose of it.[12]

Municipal solid waste is the subset of solid waste, as defined above, generated by the community.[13] The sources of municipal solid waste can include residential, commercial, institutional, construction and demolition, municipal services (such as street cleaning), and treatment sites (such as municipal solid waste incinerators).[14]

Solid Waste composition is largely affected by two factors: (1) income level of the country; and (2) extent of industrialization.[15] Income level affects the main ingredients in solid waste, particularly the level of packaging (i.e., paper, plastic, carton, cans, and bottles).[16]

Types of Solid Wastes

There are two basic sources of solid wastes: non-municipal and municipal. Non-municipal solid waste is the discarded solid material from industry, agriculture, mining, and oil and gas production.[17] Municipal solid waste is made up of discarded solid materials from residences, businesses, and city buildings. It consists of materials from plastics to food scraps. The most common waste product is paper (about 40 percent of the total).[18]

Some other scholars have classified Solid Wastes differently. According to them Solid waste can be classified into different types depending on their source:

(a) Household waste generally classified as municipal waste,

(b) Industrial waste as hazardous waste, and

(c) Biomedical waste as hospital waste or infectious waste.[19]

Municipal solid waste: is produced as a result of economic productivity and consumption.[20] Cities with higher incomes produce more waste per capita and per employee, and their wastes have higher portions of packaging materials and recyclable wastes.[21] In low-income cities, there is less commercial and industrial activity, as well as less institutional activity, thus resulting in lower waste generation rates.[22]

Municipal solid waste consists of household waste, construction and demolition debris, sanitation residue, and waste from streets.[23] This garbage is generated mainly from residential and commercial complexes.[24] In 1947 cities and towns in India generated an estimated 6 million tones of solid waste; in 1997 it was about 48 million tones. More than 25% of the municipal solid waste is not collected at all; 70% of the Indian cities lack adequate capacity to transport it and there are no sanitary landfills to dispose of the waste.[25] The existing landfills are neither well equipped nor well managed and are not lined properly to protect against contamination of soil and groundwater.[26]

Over the last few years, the consumer market has grown rapidly leading to products being packed in cans, aluminium foils, plastics, and other such non- biodegradable items that cause incalculable harm to the environment.[27] In India, some municipal areas have banned the use of plastics and they seem to have achieved success. For example, today one will not see a single piece of plastic in the entire district of Ladakh where the local authorities imposed a ban on plastics in 1998. One positive note is that in many large cities, shops have begun packing items in reusable or biodegradable bags. Certain biodegradable items can also be composted and reused. In fact proper handling of the biodegradable waste will considerably lessen the burden of solid waste that each city has to tackle.[28]

The composition of the municipal wastes can vary from region to region and from season to season.[29] Food waste, which includes animal and vegetable wastes resulting from the preparation and consumption of food, is commonly known as garbage.[30]

Some solid wastes are detrimental to the health and well-being of humans. These materials are classified as hazardous wastes.[31] Hazardous wastes are defined as materials which are toxic, carcinogenic (cause cancer), mutagenic (cause DNA mutations), teratogenic (cause birth defects), highly flammable, corrosive or explosive.[32] Although hazardous wastes in the United States are supposedly regulated, some obviously hazardous solid wastes are excluded from strict regulation; these include: mining, hazardous household and small business wastes.[33]

Hazardous waste: Industrial and hospital waste is considered hazardous as they may contain toxic substances.[34] Hazardous wastes could be highly toxic to humans, animals, and plants; are corrosive, highly inflammable, or explosive; and react when exposed to certain things e.g. gases.[35] India generates around 7 million tonnes of hazardous wastes every year, most of which is concentrated in four states: Andhra Pradesh, Bihar, Uttar Pradesh, and Tamil Nadu.[36]

Household wastes that can be categorized as hazardous waste include old batteries, shoe polish, paint tins, old medicines, and medicine bottles.[37]

Hospital waste contaminated by chemicals used in hospitals is considered hazardous. These chemicals include formaldehyde and phenols, which are used as disinfectants, and mercury, which is used in thermometers or equipment that measure blood pressure.[38] Most hospitals in India do not have proper disposal facilities for these hazardous wastes.

In the industrial sector, the major generators of hazardous waste are the metal, chemical, paper, pesticide, dye, refining, and rubber goods industries.[39] Direct exposure to chemicals in hazardous waste such as mercury and cyanide can be fatal.[40]

Hospital waste: Hospital waste is generated during the diagnosis, treatment, or immunization of human beings or animals or in research activities in these fields or in the production or testing of biologicals.[41] These are in the form of disposable syringes, swabs, bandages, body fluids, human excreta, etc.[42] This waste is highly infectious and can be a serious threat to human health if not managed in a scientific and discriminate manner.[43]

Surveys carried out by various agencies show that the health care establishments in India are not giving due attention to their waste management.[44] After the notification of the Bio-medical Waste (Handling and Management) Rules, 1998, these establishments are slowly streamlining the process of waste segregation, collection, treatment, and disposal. Many of the larger hospitals have either installed the treatment facilities or are in the process of doing so.[45]

Plastic Bags as Solid Wastes

Plastic with its exclusive qualities of being light yet strong and economical, has invaded every aspect of our day-to-day life.[46] Once hailed as a 'wonder material', plastic is now a serious worldwide environmental and health concern, essentially due to its non- biodegradable nature.[47]

In India, the plastic industry is growing phenomenally. But the good news is that along with a growth in the use, a country-wide network for collection of plastic waste through rag pickers, waste collectors and waste dealers and recycling enterprises has sprung all over the country over the last decade or so.[48] More than 50% of the plastic waste generated in the country is recycled and used in the manufacture of various plastic products.

Conventional plastics have been associated with reproductive problems in both wildlife and humans. Studies have shown a decline in human sperm count and quality, genital abnormalities and a rise in the incidence of breast cancer.[49] Dioxin a highly carcinogenic and toxic by-product of the manufacturing process of plastics is one of the chemicals believed to be passed on through breast milk to the nursing infant.[50] Burning of plastics, especially PVC releases this dioxin and also furan into the atmosphere.[51] Thus, conventional plastics, right from their manufacture to their disposal are a major problem to the environment.

Careless disposal of plastic bags chokes drains, blocks the porosity of the soil and causes problems for groundwater recharge.[52] Plastic disturbs the soil microbe activity, and once ingested, can kill animals. Plastic bags can also contaminate foodstuffs due to leaching of toxic dyes and transfer of pathogens.[53] In fact, a major portion of the plastic bags i.e. approximately 60-80% of the plastic waste generated in India is collected and segregated to be recycled. The rest remains strewn on the ground, littered around in open drains, or in unmanaged garbage dumps.[54]

The plastic industry in the developed world has realized the need of environmentally acceptable modes for recycling plastics wastes and has set out targets and missions to acknowledge the need for plastics to conform to certain guidelines/standards and code of conduct for its use.[55]

Designing eco-friendly, bio- degradable plastics is the need of the hour. Though partially biodegradable plastics have been developed and used, completely biodegradable plastics based on renewable starch rather than petrochemicals have only recently been developed and are in the early stages of commercialization.[56]

Solid waste management

'Solid waste management is a polite term for garbage management. As long as humans have been living in settled communities, solid waste, or garbage, has been an issue, and modern societies generate far more solid waste than early humans ever did. Solid waste management is a system for handling all of this garbage; municipal waste collection is solid waste management, as are recycling programs, dumps, and incinerators[57]. Earlier in ancient times solid waste management consisted of digging pits and throwing garbage into them. When cities began to be more concentrated, however, solid waste management became a serious issue. So in response, many cities started to set up municipal garbage collection, in the form of rag and bone men who would buy useful garbage from people and recycle it, or waste collection teams which would dispose of unusable garbage.[58] For most industrialized nations today, solid waste management is a multibillion dollar business which is also crucial to survival. Garbage collection agencies remove tons of garbage yearly and sort it for recycling or ultimate disposal. Most cities require citizens to pay for waste collection, while rural areas have dumps and recycling facilities for citizens to bring their garbage to[59] . The end goal is a reduction of the amount of garbage clogging the streets and polluting the environment, whether that garbage is disposed of or recycled into something useful [60].

There are many types of solid waste which need to be dealt with. The first is recyclable waste, objects which are useful, but no longer wanted. Solid waste management includes the construction of facilities to recycle these goods, which include scrap metal, glass, cans, paper, plastics, wood, and similar materials.[61] Another category is toxic waste; waste which could potentially contaminate the environment, meaning that it needs to be handled with care. This category includes electronic waste, a growing problem in many industrialized nations. Next is green waste such as compost and yard clippings.[62] People with land can compost their own green waste, and many cities collect it separately, so that the green waste can be composted and returned to the earth.

"Management of solid waste is one of the most neglected areas of development in India."[63]. In India, Solid waste of more than 30 million tonnes annually ends up on the outskirts of major urban locations, where the waste is burned, or produces methane while rotting, or results in epidemics and has an adverse effect on rag pickers and sanitary workers[64]. About 0.1 million tones of municipal solid waste are generated in India and Only 5% of the collected municipal waste is composted. Fifty Seven per cent of Delhi's waste is dumped in Yamuna[65].

> *"Solid Waste management is increasingly becoming a critical issue for municipal issue for municipal authorities. Central and the state governments are supportive of local efforts to improve MSW (Municipal Solid Waste) management but this is essentially a municipal function and it is at this level that challenges have to be addressed."* [66]

Health impacts of solid waste management

All activities in solid waste management involve risk, either to the worker directly involved, or to the nearby resident. Risks occur at every step in the process, from the point where residents handle wastes in the home for collection or recycling, to the point of ultimate disposal.[67] High levels of contamination of air, soil and water in a few well publicized situations have led to widespread unease about the potential health effects of waste management processes, particularly within communities living in the proximity to relevant sites.[68]

The group at risk from the unscientific disposal of solid waste include – the population in areas where there is no proper waste disposal method, especially the pre-school children; waste workers; and workers in facilities producing toxic and infectious material.[69] Other high-risk group includes population living close to a waste dump and those, whose water supply has become contaminated either due to waste dumping or leakage from landfill sites. Uncollected solid waste also increases risk of injury, and infection.[70]

I. Occupational and environmental health risks

Workers and waste pickers handling solid waste throughout the world are exposed to occupational health and accident risks related to the content of the materials they are handling, emissions from those materials, and the equipment being used.[71]

Much of the health literature on the toxicity of the individual substances relates to occupational or accidental exposure and thus generally to higher levels of exposure than those expected from waste disposal methods.[72] Many of the substances, such as cadmium, arsenic, chromium, nickel, dioxins are considered to be carcinogenic, based on animal studies or studies of people exposed to high levels.[73] In addition to carcinogenicity, many of these substances can produce other toxic effects (depending on exposure level and duration) on the central nervous system, liver, kidneys, heart, lungs, skin, reproduction, *etc.*[74] For other pollutants such as SO_2 and PM_{10}, air pollution studies have indicated that there may be effects on morbidity and mortality at background levels of exposure, particularly in susceptible groups such as the elderly.[75] Chemicals such as dioxins and organochlorines may be lipophilic and accumulate in fat-rich tissues and have been associated with reproductive or endocrine-disrupting endpoints.[76]

Landfill sites

One of the mostly widely known and publicized landfill sites is that of Love Canal in New York State. By the mid 1970s, chemicals leaking from the site were detected in local streams, sewers, soil and indoor air of houses.[77] Since then there have been many studies of populations living near landfill sites, frequently carried out near one specific site in response to public concern.[78] In addition, several large studies have been carried out investigating health outcomes near hundreds of sites.[79]

(i) Birth defects and reproductive disorders

Reproductive effects associated with landfill sites have been extensively researched and include low birth weight (less than 2500 g), fetal and infant mortality, spontaneous abortion, and the occurrence of birth defects.[80] Vianna and Polan[81] and Goldman *et al*[82] both found increased incidence of low birth weight in the populations around the Love Canal site. A similar increase in the proportion of low birth weight babies was found in those living within a radius of 1 km of the Lipari Landfill in New Jersey, particularly in 1971–75 following a period of heavy pollution of streams and a nearby lake from leachate from the site[83]. Trends in low birth weight and neonatal deaths were found to correspond closely with time and quantities of dumping at a large hazardous waste disposal site in California, with significantly lower birth weights in exposed areas than control areas during the periods of heaviest dumping[84].

The results from these single site studies for low birth weight contrast with results from two large multiple site case-control studies in the USA[85]. An interesting finding from this study was that 80% of the population in Great Britain lives within 2 km of an operating or closed landfill site.[86]

In the two US multiple site studies, one[87] found a small increase (1.5-fold) in heart and circulatory malformations but no increased risk for other malformations. The other[88] found no association, although the response to the questionnaire used to collect data was relatively poor (63%) and it is unclear how congenital malformations were defined. The UK study[89] found significantly elevated risks for several defects, including neural tube defects, hypospadias and epispadias, abdominal wall defects and surgical correction of gastroschisis and exomphalos, etc. A similar finding was also reported in the analysis of congenital malformation rates among the population living near the Welsh landfill of Nant-y-Gwyddon where nearly double the risk was found in exposed areas both before and after the site opened.[90] Risk was raised for neural-tube defects, malformations of the cardiac septa and anomalies of great arteries and veins[91]. A study by the same group showed similar increases in chromosomal anomalies, even after adjustment for maternal age[92].

A similar study was carried out in New York State but also attempted to investigate associations with off-site migration of chemicals and certain categories of chemicals present at the sites[93]. A small (12%) statistically significant risk of congenital malformations was associated with maternal proximity to a site which increased with off-site chemical leaks.[94]

(ii) Cancer

Several geographical comparison studies have investigated cancer mortality and incidence around waste sites. Increased frequency of cancers in countries containing hazardous waste sites was found in two US studies[95], particularly for gastrointestinal, oesophageal, stomach, colon and rectal cancer. No increase in cancer rates or the frequency of chromosome changes was found in relation to the Love Canal site[96]. Two reports[97] of cancer incidence among persons living near the Miron Quarry site, the third largest in North America found increased incidence of cancers of the liver, kidney, pancreas and non-Hodgkin's lymphomas.

(iii) Studies of self-reported health symptoms

Many of the studies investigating health outcomes other than birth defects and reproductive orders and cancers have been community health surveys and have relied on the self-reporting of symptoms through interviews or questionnaires[98]. The health problems investigated include respiratory symptoms, irritation of the skin, nose and eyes, gastrointestinal problems, fatigue, headaches, psychological problems and allergies.[99]

Incineration

Evaluation of the potential health effects of the large number of pollutants which can be produced by waste incineration can be approached by assessing the effects of individual pollutants[100] or through more general studies of community residents[101] and incinerator workers.

(i) Individual pollutants

From the health aspect, the most important pollutants associated with incineration are particles, acidic gases, aerosols, metals and organic compounds. There is an extensive research literature on both the acute and chronic effects of particles.[102] Effects appear to be more severe in susceptible groups such as children, the elderly, or those with chronic conditions such as asthma or pre-existing cardiovascular diseases[103].

Although less well established, results from large US cohort studies suggest that long-term exposure to low concentrations is associated with chronic health effects such as increased rates of bronchitis and reduced lung function[104], shortened life span, and elevated rates of respiratory symptoms and lung cancer[105].

Studies of outdoor levels of NO_2 and health effects may be hampered by difficulties in separating the effects of the various components of ambient pollution.[106] Less equivocal are results relating to acidic gases, in particular sulphur dioxide (SO_2), for which there is also an extensive literature[107]. Asthmatics appear to be particularly sensitive to the effects of SO_2 on lung function, although the concentrations at which these occur vary between studies.[108]

Metals associated with incinerator emissions at various levels and *via* various media and exposure pathways have all been shown to cause a range of carcinogenic and non-carcinogenic health effects.[109]

The organic compounds which have received the most attention relating to incineration are dioxins and PCBs, partly because of their ability to accumulate in the body.[110] High levels of dioxin exposure found in workplaces and after accidents such as that at Seveso have caused chloracne and an increase in cardiovascular disease.[111]

(ii) Health effects in communities

Most of the studies of communities living near incinerators have assessed exposure using some measure of distance from the site or an estimate of areas at most risk from emissions. Studies focusing on a single waste incinerator suggested some relationship between distance from the site and mortality or incidence from some cancers.[112] A series of studies in the UK[113] of multiple sites compared observed cancer incidence rates in bands of increasing distance from each incinerator with rates based on national data. A study of residence near MSW incinerators found statistically increasing risk with increasing proximity for all cancers and for colorectal, lung, liver and stomach cancers, although there was evidence of residual confounding for all cancers, stomach and lung.[114] Because of the substantial level of misdiagnosis which can occur among registrations and death certificates for liver cancer, the authors carried out a histological review of the cases and reanalysis for this disease. Reduced estimates of excess incidence of primary liver cancer were reported[115].

(iii) Worker populations

There is a large workforce employed in waste collection, sorting and disposal. Workers may be exposed to the same potential hazards as the general population, although the amount of exposure and risk may differ.[116] The type of work varies between waste management options with some, such as landfill and incineration, being more automated than others, such as waste collection, sorting and recycling.[117] The incidence of occupational accidents in waste collection workers has been found to be higher than the general workforce[118]. It has been suggested that increased exposure to bio-aerosols and volatile compounds may lead to elevated incidence of work-related respiratory gastrointestinal and skin problems in waste collections compared to the general workforce.[119]

In addition to these, dust levels have been found to be high at refuse transfer stations and incinerators.[120] However, reduced mortality from lung cancer was found for Italian incinerator workers[121].

A small cross-sectional study of 58 compost workers[122] found significantly increased antibody concentrations against fungi and actinomycetes.

Remediation of waste and waste sites is an expanding activity, particularly the mediation of hazardous or toxic waste. The health of the workforce involved in this is an important issue[123]. An assessment of the risk of occupational fatalities associated with hazardous waste site remediation estimated that the fatality risks to remediation workers were orders of magnitude greater than human cancer risk, and that truck drivers and labourers were particularly at risk[124].

II. Role of Plastics

The unhygienic use and disposal of plastics and its effects on human health has become a matter of concern. Coloured plastics are harmful as their pigment contains heavy metals that are highly toxic.[125] Some of the harmful metals found in plastics are copper, lead, chromium, cobalt, selenium, and cadmium.[126] In most industrialized countries, colour plastics have been legally banned. In India, the Government of Himachal Pradesh has banned the use of plastics and so has Ladakh district. Other states should emulate their example.[127]

III. Preventive and Interim Measures

A range of prevention and control measures could potentially ameliorate the diverse risks posed by using urban organic solid waste.[128] Since there is little immediate prospect of effective intervention in the many informal activities, we must take what comfort we can from positive trends of 'self-help' in this field, and seek the easiest avenues for introducing information and low-cost technology at the urban and community level.[129]

Proper methods of waste disposal have to be undertaken to ensure that it does not affect the environment around the area or cause health hazards to the people living there.[130]

At the household-level proper segregation of waste has to be done and it should be ensured that all organic matter is kept aside for composting, which is undoubtedly the best method for the correct disposal of this segment of the waste.[131] In fact, the organic part of the waste that is generated decomposes more easily, attracts insects and causes disease. Organic waste can be composted and then used as a fertilizer.[132]

Every city needs to implement record keeping on the health of its solid waste workers, including the informal waste pickers/ recyclers.[133] Rather than having open access of waste pickers to solid waste disposal sites, they should be registered, carry photo identification at all times, and participate in a regular vaccination and health examination program.[134] Local medical schools and occupational health institutes need to be encouraged to study the health of solid waste workers in comparison with appropriate baseline control populations, since true epidemiological data is lacking for this sector.[135]

Cities that use private companies to collect, process or disposal of their solid wastes need to include special clauses in each contract, requiring workers health and safety protective measures, including annual medical check-ups and vaccinations.[136]

A reduction in the quantity of solid wastes being generated is a primary way of reducing environmental and occupational health effects of waste management.[137] Education is needed to promote manufacturing technologies and consumer practices that generate less waste.[138]

From a health and injury perspective, waste picking should be prevented within sanitary landfills.[139] Until new modern landfills can be implemented and existing open dumps closed, existing communities of waste pickers need assistance to make their recycling efforts safer, increase their sorting and waste processing productivity, and network with potential buyers of secondary materials.[140] Technical assistance needs to be provided to help waste pickers transition from independent informal sector workers to organized enterprises, cooperatives, or economic groups.[141]

Special wastes, such as infectious healthcare wastes and hazardous industrial wastes, are best handled in systems separate from municipal solid waste collection and disposal.[142] The infectious wastes should be segregated at the source and placed in sealed disposable containers; and the containers should remain sealed until treated or incinerated within a closed system.[143]

Conclusions and solutions

The issue of solid waste needs much more understanding, participation, cooperation / partnership of a society's members than any other environmental issue. Great potential for such partnerships exist in traditional and indigenous local social systems. To draw out this potential, the sharing of many experiences of success and failure, and their contributing factors, among local societies across borders would be very beneficial.

So to maximize waste reduction in order to minimize disposal cannot be done unless residents, businesses, institutions, and government take increased responsibility for reducing, reusing, and recycling waste[144]. Manufacturers also must change their role by taking more responsibility for the products they produce so that they contain less toxic, create less waste, and are easier to recycle. To effect these behavioral changes, it will be critical to forge partnerships to ensure that everyone takes responsibility for their share. "Both knowledge and attitudes are assumed to be important for changing human actions toward the environment"[145].

So Citizens can do a lot to reduce the amount of waste they dispose of. They can[146]:

- Learn about the recycling programs that are available to them
- Take advantage of the recycling programs their municipalities offer (and work with their local officials to expand these programs for their communities[147]
- Buy "green" products that are manufactured with no or lower levels of toxic substances and that are designed to be recycled at the end of their useful life.

- Support legislation that will broaden the recycling programs currently available and establish Product Stewardship initiatives with manufacturers[148]

Endnotes

1. Diaz, L. et al. *Solid Waste Management, Volume 2.* UNEP/Earthprint, 2006
2. Ray, A. "Waste management in developing Asia: Can trade and cooperation help?" *The Journal of Environment & Development* 17.1 (2008): 3-25
3. Available at http://en.wikipedia.org/wiki/Waste
4. Baker, Elaine et al. "Vital Waste Graphics." United Nations Environment Program and Grid-Arendal, 2004.
5. "Glossary of Statistical Terms." 2003. OECD. 12 October.
6. Pongrácz E & Pohjola VJ. "Re-defining waste, the concept of ownership and the roles of waste management" Resources Conservation & Recycling. 40.2 (2004): 141-153
7. Environment Agency. **Waste** Statistics for Engl**and and** Wales 1998–99. Environment Agency
8. Available at http://bmb.oxfordjournals.org/cgi/content/full/68/1/183
9. Mechanical Biological Treatment Welsh Assembly (2005) Mechanical Biological Treatment, Environment Countryside and Planning Website, Welsh Assembly
10. P.L. 62-701, Section 102
11. *Id.*
12. "Muck and brass: The waste business smells of money." *The Economist.* 2009 02 28. pp. 10-12
13. Wilson, D.C.; Velis, C.; Cheeseman, C. "Role of informal sector recycling in waste management in developing countries." *Habitat International* 30 (2006): 797-808
14. Englehardt James, Fleming E. Laura, Bean A. Judy, AN Huren, John Nicollette, Rogers Jeff, Danits Melissa, "Solid Waste Management Health and Safety Risks: Epidemiology and

Assessment to Support Risk Reduction", University of Miami, March, 2000

15. Jacbos CB. *EPA Landfill Methane Outreach Program*. Proceedings of the 17th Annual Landfill Gas Symposium (Long Beach, California) of the Solid Waste Management Association of North America. March 1994: 1-11

16. Kerrison R. *Bad Smell in 'Cleaned-Up' Garbage-Hauling Business*. New York Post. September 28, 1997

17. International Waste Activities." 2003. U.S. Environmental Protection Agency. 12 Oct 2009

18. William D. Ruckelshaus, The Politics of Waste Disposal, Wall St. J., Sept.5, 1989, at A18

19. Improving Recycling Markets." OECD Environment Program. Paris: OECD, 2006

20. Klinck BA. A Groundwater Hazard Assessment Scheme for Solid Waste Disposal: Summary Report. Technical Report WC/96/17. British Geological Survey, Overseas Geology Series. Prepared for the Overseas Development Administration, London, UK. 1996

21. Kocasoy G, Curi K. *Umraniye-Hekimbasi Open Dump Accident*. Waste Management and Research. Journal of the International Solid Waste Association. 1995; Vol. 13: 305-14.Q34

22. Dok H, Vrljheld M, Armstrong B., Abramsky L, et al. *Risk of Congenital Anomalies Near Hazardous-Waste Landfill Sites in Europe: the EUROHAZCON Study*. The LANCET. Vol. 352. London, UK. August 8, 1998

23. Gladding T. *Health and Safety Aspects of Household Waste Recycling*. Waste Management. July 1993

24. Huisman M. *The Position of Waste Pickers in Solid Waste Management*. In: Baud I, Schenk H. Solid Waste Management: Modes, Assessments, Appraisals, and Linkages in Bangalore, Manohar, New Delhi. 1994: 46-104

25. Available at www.indianstat.org

26. Gray T. *Rat Race, 1993*. World Health. May-June 1993; Vol. 46, No. 3: 28

27. Poulsen, O.M., Breum, N.O., Ebbehoj, N., et al. "Collection of domestic waste. Review of occupational health problems and their possible causes," Science of the Total Environment 170;1-19 (1995a)

28. Schoenborn, C.A.: for National Center for Health Statistics. Health Promotion and Disease Prevention: United States, 1985. Vital and Health Statistics. Series 10, Number 163. DHHS Publishing Number

(PHS) 88-1591, Public Health Service. Washington, DC: U.S. Government Printing Office, 1988

29. Kerrison R. *Bad Smell in 'Cleaned-Up' Garbage-Hauling Business*. New York Post. September28, 1997

30. Konnoth N. The Forum for Environmental Concern. Your Clean City at Whose Cost: A Study on the Working Conditions and Occupation Hazards at the Dumping Sites of Bombay. 1-56, plus annexes

31. German Agency for Technical Cooperation. Scavenger Activities and Health Hazards to Scavengers. May 1986: 1-38

32. Goldberg MS, Al-Homsi N, et.al. *Incidence of Cancer Among Persons Living Near a Municipal Solid Waste Landfill Site in Montreal, Quebec*. Archives of Environmental Health. November/December 1995; Vol. 50, No. 6: 416-24

33. Hovde DC, Stanton AC, et.al. *Methane Emissions from a Landfill Measured by Eddy Correlation Using a Fast Response Diode Laser Sensor*. Journal of Atmospheric Chemistry. February 1995; Vol.20, No. 2: 141-63

34. Kanitz S, Franco Y, et.al. *Sanitary Landfilling: Occupational and Health Hazards*. Proceedings of Sardinia 91, Third International Landfill Symposium, Cagliari, Italy. October 1991

35. Tchobanoglous, G., Theisen, G., Vigil, S. Integrated Solid Waste Management: Engineering Principles and Management Issues. New York, NY: McGraw-Hill Inc, 1993

36. Available at www.indianstat.org

37. Johnson BL, DeRosa CT. *The Toxicological Hazard of Superfund Hazardous Waste Sites*. Reviews on Environmental Health, Freund Publishing House, Vol. 12, No. 4, pp.235-251, 1997

38. Feachem RG, Bradley DJ, et.al. Sanitation and Disease: Health Aspects of Excreta and Wastewater Management. John Wiley & Sons, Chichester, UK. 1983. 1-501

39. Elliott P, Shaddick G, et.al. *Cancer Incidence near Municipal Solid Waste Incinerators in Great Britain*. British Journal of Cancer. 1996: 702-10

40. Dok H, Vrljheld M, Armstrong B., Abramsky L, et al. *Risk of Congenital Anomalies Near Hazardous-Waste Landfill Sites in Europe: the EUROHAZCON Study*. The LANCET. Vol. 352. London, UK. August 8, 1998

41. Cullen MR, Cheerniack MG, Rosenstock L. *Medical Progress: Occupational Health.* The New England Journal of Medicine. March 1, 1990: 594-601

42. Cucu M, Iacob I, et. al. National Integrated Programme on Environmental and Health in Romania, 1993 - 1996: Final Report on the Study of Acute Environmental Health Problems from Waste Management in Romania. Institute of Hygiene, Public Health, Health Services and Management - Bucharest. Sponsored by the World Health Organization, Regional Office for Europe, Nancy Project Office, France. 1996

43. de Coura Cuentro S, Dji Malla Gadji. The Collection and Management of Household Garbage. The Poor Die Young: Housing and Health in Third World Cities, edited by Hardoy JE, Caincross S, Satterhwaite D. Earthscan, London, UK. 1990

44. Available at www.indianstat.org

45. Cullen MR, Cheerniack MG, Rosenstock L. *Medical Progress: Occupational Health.* The New England Journal of Medicine. March 1, 1990: 594-601

46. N. Y. Comp. Codes R. & Regs. tit. 10, 19.1- .5 (1962)

47. The Nelson A. Rockefeller Institute of Government, The Department of Environmental Conservation: A 25th Anniversary Review 65 (1996)

48. Dennis Hevesi, The Long History of A Toxic- Waste Nightmare, N.Y. Times, Sept.28, 1988, at B4

49. Philip S. Gutis, For Trash Barge Crew, Empty Days and Flies, N.Y. Times, May 4, 1987, at A1

50. Barbara Hogan, All Baled up and No Place to Go, Conservationalist, Jan.- Feb., 1988, at 37

51. Major Chnages due For Solid Waste Regulations, Wasteline (New York Sate Dept. of Envtl. Conservation, Albany, N.Y.), Nov., 1987

52. Division of Solid Waste, New York State Dept. of Envtl. Conservation, New York State Solid Management Plan 1989-1990

53. Erica L. Heintz, Where will the Garbage Go?, in New York State Assembly: Legislative Commission on Solid Waste Management 32 (April 1995)

54 .Memishi, R. "Hazards of the Waste Industry". Waste News, Crain Communications, Inc., Detroit, 4: 1-24 (1998)

55. Midtgard, U. & Poulsen, O. M. "Occupational Safety and Health in Waste Collection and Recycling; The CORE Research Program." Ann. Agric. Environ. Med. 4: 21-26 (1997)

56. Florida Department of Environmental Protection, Bureau of Solid and Hazardous Waste Solid Waste Management in Florida, 2000

57. David C. Wilson, Waste Management § 1.3.1, at 9-13 (1981).See also, http://www.wisegeek.com/what-is-solid-waste-management.html

58. Raymond Zanney, Ohio's Solid Waste Disposal: A Major Challenge, in Environment Cleveland 56, 57-60 (David Barnhizer ed., 1990) (discussing the concept of integrated waste management and alternative methods of solid waste disposal).

59. John F. Crawford & Paul G. Smith, Landfill Technology § 2.1 (1985).

60. *Id.*

61. Edward Byers, Now Entering the Age of NIMBY? , Waste Age , Jan. 1990, at 36, 36-37

62. James T. O'Reilly , State & Local Government Solid Waste Management § 8:07 (1991).

63. Raymond Zanney, Ohio's Solid Waste Disposal: A Major Challenge, in Environment Cleveland 56, 57-60 (David Barnhizer ed., 1990). See also, http://www.chillibreeze.com/articles_various/Solid-Waste-Management.asp

64. U.S. Environmental Protection Agency, The Solid Waste Dilemma: An Agenda For Action 8, 14-15 (1989) (describing the NIMBY syndrome) [hereinafter U.S. EPA]. For a general but comprehensive analysis of NIMBY, see Denis J. Brion, An Essay on Lulu, Nimby, and the Problem of Distributive Justice , 15 B.C. Envtl. Aff. L. Rev. 437 (1988) (tracing the roots of NIMBY to the law's failure adequately to protect social values).

65. http://www.organicsolutions.in/Solid-Waste-Treatment.htm

66. *Id.* See also, Bernd Holznagel, Negotiation and Mediation: The Newest Approach to Hazardous Waste Facility Siting , 13 B.C. Envtl. Aff. L. Rev. 329 (1986).

67. Sandra Cointreau, Occupational and Environmental Health Issues of Solid Waste Management, The World Bank, Washington D.C.

68. Kocasoy G, Curi K. *Umraniye-Hekimbasi Open Dump Accident.* Waste Management and Research. Journal of the International Solid Waste Association. 1995; Vol. 13: 305-14.Q34

69. Johnke B, Stelzner E. *Results of the German Dioxin Measurement Programme at MSW Incinerators*. Waste Management and Research. Journal of the International Solid Waste Association. 1992; Vol.10, No. 4: 345-55

70. Jacbos CB. *EPA Landfill Methane Outreach Program*. Proceedings of the 17th Annual Landfill Gas Symposium (Long Beach, California) of the Solid Waste Management Association of North America. March 1994: 1-11

7. Environment Agency. ***Waste** Statistics for England **and** Wales 1998–99*. Environment Agency

72. Coen LA, Shaw GM, et. al., *Maternal Residential Proximity to Hazardous Waste Sites and Risk for Selected Congenital Malformations*. Epidemiology, Volume 8, Number 4. July 1997

73. Newsletter of the U.S. Department of Health and Human Services, Agency for Toxic Substances and Disease Registry. Volume 8, Number 2. Winter 1998

74. Hovde DC, Stanton AC, et.al. *Methane Emissions from a Landfill Measured by Eddy Correlation Using a Fast Response Diode Laser Sensor*. Journal of Atmospheric Chemistry. February 1975; Vol. 20, No. 2: 141-63

75. Henriksen AZ, et al. *Severe Gastroenteritis after Infection with Vibrio cholerae non-01*. Tidsskrift Norsk Laegeforening. October 1993; Vol. 113, No. 24: 3017-18

76. Indorf AW. Occupational Health Hazards of Recycling Workers at the Bonanza Landfill in Venezuela. Masters Thesis, Clark University, Worcester, Massachusetts, USA. 1994

77. Berry, M., Bove, F., New Jersey Department of Health, USA, *Birth weight reduction associated with residence near a hazardous wate landfill*, Environmental Health Perspective, 105(8):856-61, August 1993

78. Bonnerjee A. Death in the Dumps. The Statesman. July 12, 1982: 4

79. Carotti AA, Smith RA. Gaseous Emissions from Municipal Incinerators. US Environmental Protection Agency, Washington DC, USA. 1974: 1-61

80. Cimino JA. *Health and Safety in the Solid Waste Industry*. American Journal of Public Health. January 1975; Vol.65, No. 1: 38-46

81. Vianna NJ, Polan AK. Incidence of low birthweight among Love Canal residents. *Science* 1984; 226: 1217–9

82. Goldman LR, Paigen B, Magnant MM, High**land** JH. Low birthweight, prematurity **and** birth defects in children living near

the hazardous **waste** site, Love Canal. *Hazard* ***Waste*** *Hazard Mater* 1985; 2: 209–23

83. Berry M, Bove F. Birthweight reduction association with residence near a hazardous waste landfill. *Environ Health Perspect* 1997; 105: 856–61

84. Kharazi M, Von Behren J, Smith M, Lomas T, Armstrong M, Broadwin R *et al.* A community based study of adverse pregnancy outcomes near a large hazardous waste landfill in California. *Toxicol Ind Health* 1997; 13: 299–310

85. Shaw GM, Schulman J, Frisch JD, Cummins SK, Harris JA. Congenital malformations and birthweight in areas with potential environmental contamination. *Arch Environ Health* 1994; 47: 147–54; Sosniak WA, Kaye WE, Gomez TM. Data linkage to explore the risk of low birthweight associated with maternal proximity to hazardous waste sites from the National Priorities list. *Arch Environ Health* 1994; 49: 251–5

86. Elliott P, Briggs D, Morris S, de Hoogh C, Hurt C, Jensen TK *et al.* Risk of adverse outcomes in populations living near landfill sites. *BMJ* 2000; 363–8

87. Shaw GM, Schulman J, Frisch JD, Cummins SK, Harris JA. Congenital malformations and birthweight in areas with potential environmental contamination. *Arch Environ Health* 1992; 47: 147–54

88. Sosniak WA, Kaye WE, Gomez TM. Data linkage to explore the risk of low birthweight associated with maternal proximity to hazardous waste sites from the National Priorities list. *Arch Environ Health* 1994; 49: 251–5

89. Elliott P, Briggs D, Morris S, de Hoogh C, Hurt C, Jensen TK *et al.* Risk of adverse outcomes in populations living near landfill sites. *BMJ* 2001; 363–8

90. Fielder HMP, Poon-King C, Palmer SR, Coleman G. Assessment of the impact on health of residents living near the Nant-y-Gwyddon landfill site: Retrospective analysis. *BMJ* 2000; 320: 19–23

91. Dolk H, Vrijheid M, Armstrong B, Abramsky L, Bianchi F, Garne E *et al.* Risk of congenital anomalies near hazardous-waste landfill sites in Europe: the EUROHAZCON study. *Lancet* 1998; 352: 423–7

92. Vrijheid M, Dolk H, Armstrong B, Abramsky L, Bianchi F, Fazarinc I *et al.* Chromosomal congenital anomalies and residence near hazardous waste landfill sites. *Lancet* 2002; 359: 320–2

93. Geshwind SA, Stowijk JAJ, Bracken M *et al.* Risk of congenital malformations associated with proximity to hazardous waste sites. *Am J Epidemiol* 1992; 135: 1197–207

94. Marshall EG, Gensburg LJ, Deres DA, Geary NS, Cayo MR. Maternal residential exposure to hazardous wastes and risk of central nervous system and musculoskeletal birth defects. *Arch Environ Health* 1997; 52: 416–25

95. Najem GR, Louria DB, Lavenhar MA, Feuerman M. Clusters of cancer mortality in New Jersey municipalities: with special reference to chemical toxic waste disposal sites and per capita income. *Int J Epidemiol* 1983; 12: 276–88; Griffith J, Duncan R, Riggan W, Peloma A. Cancer mortality in US counties with hazardous waste sites and ground water pollution. *Arch Environ Health* 1989; 44: 69–74

96. Janerich DT, Burnett WS, Feck G, Hoff M, Nasca P, Polednak AP *et al.* Cancer incidence in the Love Canal area. *Science* 1981; 212: 1404–7; Heath CW, Nadel MR, Zack MM, Chen ATL, Beuder MA *et al.* Cytogenetic findings in persons living near the Love Canal. *JAMA* 1984; 251: 1217–9

97. Goldberg MS, Al-Homsi N, Goulet L, Riberdy H. Incidence of cancer among persons living near a municipal solid waste landfill site in Montreal, Quebec. *Arch Environ Health* 1995; 50: 416–24; Goldberg MS, DeWar R, Desy M, Riberdy H. Risk of developing cancer relative to living near a municipal solid waste landfill site in Montreal, Quebec, Canada. *Arch Environ Health* 1999; 54: 291–6

98. Vrijheid M. Health effects of residence near hazardous waste landfill sites: a review of epidemiologic literature. *Environ Health Perspect* 2000; 108: 101–12

99. Ozonoff D, Colten ME, Cupples A, Heeren T, Schatzin A, Mangione T *et al.* Health problems reported by residents of a neighbourhood contaminated by a hazardous waste facility. *Am J Ind Med* 1987; 11: 581–97

100. IEH. *Health Effects of Waste Combustion Products*. Leicester: Institute for Environment and Health, 1997; Committee on Health Effects of Waste Incineration. *Waste Incineration and Public Health*. Washington, DC: National Academy Press, 2000

101. Hu S, Shy CM. Health effects of waste incineration: a review of epidemiologic studies. *J Air Waste Manage Assoc* 2001; 51: 1100–9

102. Dockery DW, Pope III CA. Acute respiratory effects of particulate air pollution. *Annu Rev Public Health* 1994; 15: 107–32; Katsouyanni

K, Touloumi G, Spix C, Schwartz J, Balducci F, Medina S *et al.* Short term effects of ambient sulphur dioxide and particulate matter on mortality in 12 European cities: Results from time series data from the APHEA project. *BMJ* 1997; 314: 1658–63

103. Zanobetti A, Schwartz J, Gold D. Are there sensitive subgroups for the effects of airborne particles? *Environ Health Perspect* 2000; 108: 841–5

104. WHO. *Air Quality Guidelines for Europe* (European Series No 91). Copenhagen: World Health Organization Regional Office for Europe, 2000

105. EPAQPS. *Airborne Particles.* London: The Stationery Office, 2001

106. Dockery DW, Pope III CA. Acute respiratory effects of particulate air pollution. *Annu Rev Public Health* 1985; 15: 107–32

107. Zanobetti A, Schwartz J, Gold D. Are there sensitive subgroups for the effects of airborne particles? *Environ Health Perspect* 2002; 108: 841–5

108. Alabaster GP. Policy Issues for the Promotion of Waste Recycling and Reuse in Developing Countries. United Nations Centre for Human Settlements (Habitat). Draft, unpublished. December 1995: 1-54. UNEP (Habitat), Nairobi, Kenya.

109. Asgari MM, Dubois A, Asgari M, et.al. *Association of Ambient Air Quality with Children's Lung Function in Urban and Rural Iran.* Archives of Environmental Health. Vol. 53, No. 3. May-June 1998

110. Associated Press. *Avalanche of Trash Kills Spaniard, Threatens Fish.* Toronto Star. October 13, 1995

111. Benestad C, Hagen I, et. al., *Emissions of Organic Micropollutants from Discontinuously Operated Municipal Waste Incinerators.* Waste Management & Research, Journal of the International Solid Waste Association, Volume 8. 1980

112. Upton AC. Public health aspects of toxic chemical disposal sites. *Annu Rev Public Health* 1989; 10: 1–25

113. Elliott P, Hills M, Beresford J, Kleinschmidt I, Jolley D, Pattenden S *et al.* Incidence of cancers of the larynx and lung near incinerators of waste solvents and oils in Great Britain. *Lancet* 1992; 339: 854–8

114. National Research Council. *Environmental Epidemiology: Public Health and Hazardous Wastes,* vol. 1. Washington, DC: National Academy Press, 1991

115. Elliott P, Eaton N, Shaddick G, Carter R. Cancer incidence near municipal solid waste incinerators in Great Britain. Part 2:

Histopathological and case-note review of primary liver cancer cases. *Br J Cancer* 2000; 82: 1103-6

116. Elliott P, Shaddick G, Kleinschmidt I, Jolley D, Walls P, Beresford J *et al.* Cancer incidence near municipal solid waste incinerators in Great Britain. *Br J Cancer* 1996; 73: 702-10

117. Vrijheid M. Health effects of residence near hazardous waste landfill sites: a review of epidemiologic literature. *Environ Health Perspect* 2000; 108: 101-12

118. Poulsen OM, Breum NO, Ebbehøj N, Hansen AM, Ivens UI, van Lelieveld D *et al.* Collection of domestic waste. Review of occupational health problems and their possible causes. *Sci Total Environ* 1994; 170: 1-19

119. Poulsen OM, Breum NO, Ebbehøj N, Hansen AM, Ivens UI, van Lelieveld D *et al.* Sorting and recycling of domestic waste. Review of occupational health problems and their possible causes. *Sci Total Environ* 1995; 168: 33-56

120. Gustavsson P. Mortality among workers at a municipal waste incinerator. *Am J Ind Med* 1989; 15: 245-53

121. Rapiti E, Sperati A, Fano V, Dell'Orco V, Forastiere F. Mortality amongst workers at municipal waste incinerators in Rome: a retrospective cohort study. *Am J Ind Med* 1997; 31: 659-61

122. Bunger J, Antlauf-Lammers M, Schulz TG, Westphal GA, Muller MM, Ruhnan P *et al.* Health complaints and immunological makers of exposure to bioaerosols among biowaste collectors and compost workers. *Occup Environ Med* 2000; 57: 458-64

123. Johnson BL. Hazardous waste: human health effects. *Toxicol Ind Health* 1997; 13: 121-43

124. Hoskin A, Leigh J, Planek T. Estimated risk of occupational fatalities associated with hazardous waste site remediation. *Risk Anal* 1994; 14: 1011-7

125. Agency for Toxic Substances and Disease Register. *Hazardous Substances Emergency Events Surveillance.* Annual report. Atlanta: US Department of Health and Human Services, Public Health Service, 1995

126. Zmirou D, Deloraine A, Savinc P, Tillier C, Bouchanlat A, Maury N. Short term health effects of an industrial toxic waste landfill: a retrospective follow-up study in Montchanin, France. *Arch Environ Health* 1994; 49: 228-38

127. IEH. ***Health** Effects of Waste Combustion Products*. Leicester: Institute for Environment and Health, 1999

128. Nunan F. 2000. Urban organic waste markets: responding to change in Hubli-Dharwad, India. Habitat International 24:. 347-360.

129. Lardinois I & Furedy C. 1999. Source Separation of Household Waste Materials. Analysis of Case Studies from Pakistan, the Philippines, India, Brazil, Argentina, and the Netherlands. Gouda: WASTE.

130. Le TH. 1995. Urban-waste derived compost in Hanoi, Vietnam: factors affecting supply and demand. MSc thesis. Bangkok: Asian Institute of Technology.

131. Rosenberg L & Furedy C (eds). 1996. International Source Book on Environmentally Sound Technologies for Municipal Solid Waste Management. Osaka: International Environmental Technology Centre, United Nations Environment Program.

132. Asomani-Boateng R & Haight M. 1999. Assessing the performance of mechanised centralized composting plants in West Africa: the case of Teshi Nungua composting plant in Accra, Ghana. WARMER Bulletin 69 (Nov.): 4-6.

133. Furedy C & Chowdhury T. 1996. Solid Waste Re-use and Urban Agriculture: Dilemmas in Developing Countries: The Bad News and the Good News. Paper presented at Joint Congress of the Association of Collegiate Schools of Planning, and Association of European Schools of Planning, Toronto, July.

134. Blaensdorf E & Hoornweg D. 1997. The Use of Compost in Indonesia: Proposed Compost Quality Standards.Washington: World Bank, Urban Development Sector Unit.

135. Lardinois I & Furedy C. 1999. Source Separation of Household Waste Materials. Analysis of Case Studies from Pakistan, the Philippines, India, Brazil, Argentina, and the Netherlands. Gouda: WASTE.

136. Furedy C, Maclaren V & Whitney J. 1999. Re-use of Waste for Food Production in Asian cities: Health and Economic Perspectives. In: Koc M, MacRae R, Mougeot L JA & Welsh J (eds), For Hunger-proof Cities (Ottawa: International Development Research Centre), pp 136-144.

137. Hoornweg D, Thomas L & Otten L. 1999. Composting and its Applicability in Developing Countries. Urban Waste Management Working Paper Series 8. Washington, DC: World Bank.

138. Kurniawan, T.A. (2009, January 29). Excessive Solid Waste Generation and Its Impacts on the Environment. *SciTopics*.

139. Pheby D, Grey M, Giusti L, Saffron L. *Waste Management and Public Health: The State of the Evidence*. Bristol: South West Public Health Observatory, 2002

140. Fischer C, Crowe M. *Household and Municipal Waste: Comparability of Data in EEA Member Countries*. Copenhagen: European Environment Agency, 2000

141. Johnson BL, DeRosa CT. The toxicological hazards of superfund hazardous waste sites. *Rev Environ Health* 1997; 12: 235–51

142. Zmirou D, Deloraine A, Savinc P, Tillier C, Bouchanlat A, Maury N. Short term health effects of an industrial toxic waste landfill: a retrospective follow-up study in Montchanin, France. *Arch Environ Health* 1994; 49: 228–38

143. Harrad SJ, Harrison RM. *The Health Effects of the Products of Waste Combustion*. Birmingham, UK: Institute of Public and Environmental Health, University of Birmingham, 1996

144. Yeaple , Judith Anne. "Houses that Breathe" Popular Science, Mar 1991. Vol. 238, p. 80-82.

145. Donnelly, James E; Georgetown International Environmental Law Review; October 1, 2002.

146. Corson, Walter. Global Ecology Handbook: What You Can Do About the Environment Crisis; Boston: Beacon Press, 1990.

147. Harris, Mark. Embracing the Earth: Choices for Environmentally Sound Living, Chicago: The Noble Press, 1990.

148. "How to Protect Yourself from your Environment" Nat'l Wild, Aug/Sept. 1990, Vol.28, pp.30-42.

Chapter 12

Management of Solid Waste & Its Effects on Health: A Case Study of South Delhi

Arjun K Singh and Vishruti Gupta

Introduction

How can we define solid waste? In simple micro-economics terms, we can say that it's a "bad" and no one would like to consume it. In a layman's language, it is a mixture of vegetables and non- vegetable wastes in cooked and uncooked stages, leftovers, packaging of different kinds, papers, plastics, rags and other fabrics, dust, ash and a variety of combustible and non - combustible matter.

Why is Solid Waste a Matter of Concern?

It is a matter of concern because even if we want to dispose it off, it does not vanish immediately. Poor solid waste management affects health and the provision of other amenities of the society such as;

1. Transmission of disease to residents and waste collecting workers;
2. Clogging drains and sewers;

3. Visual and smell impacts;
4. Creates conditions for disease vectors such as flies, cockroaches, mosquitoes, rodent's insects etc

Literature Review

According to by Mufeed Sharholy et al (2008). Municipal solid waste management (MSWM) is one of the major environmental problems of Indian cities. Improper management of municipal solid waste (MSW) causes hazards to inhabitants. About 90% of MSW is disposed off unscientifically in open dumps and landfills, creating problems to public health and the environment. The study provides a comprehensive review of the characteristics, generation, collection and transportation, disposal and treatment technologies of MSW practiced in India. The conclusions drawn are:

- The collection bins must be appropriately designed with features like metallic containers with lids, and to have a large enough capacity to accommodate 20% more than the expected waste generation in the area, with a design for mechanical loading and un-loading, placement at appropriate locations, etc;
- Municipal authorities should maintain the storage facilities in such a manner that they do not create unhygienic and unsanitary conditions. Proper maintenance of the MSW transportation vehicles must be conducted, and the Dumper Placer should replace the old transportation vehicles in a phased manner;
- Currently, at the level of waste generation and collection, there is no source segregation of compostable waste from the other non-biodegradable and recyclable waste. Proper segregation would lead to better options and opportunities for scientific disposal of waste;
- Most of the MSW in India is dumped on land in an uncontrolled manner. Such inadequate disposal practices lead to problems that will impair human and

animal health and result in economic, environmental and biological losses;

- An open dump or an uncontrolled waste disposal area should be rehabilitated. It is advisable to move from open dumping to sanitary landfilling in a phased manner. Landfilling should be restricted to non-biodegradable, inert waste and other waste that are not suitable either for recycling or for biological processing;

The current regulations (MSWM rules, 2000) are very stringent. Norms have been developed to ensure a proper MSWM system. There is a large gap between policy prescriptions and its implementation. Another study by Papiya Sarkar (2003) emphasizes on SWM in Delhi. On an average, Delhi generates 4000 tonnes of municipal solid waste per day. A physical analysis reveals that it consists of about 32% compostable matter. The recyclable components include paper 6.6%, plastics 1.5% and metals 2.5%.It also brought out some remarkable findings about the impact of SW on various stakeholders and especially the most vulnerable section: rag pickers, scavengers and waste pickers who are engaged in this occupation. Tuberculosis, scabies, multi-system allergic disorders, asthma, respiratory infections, ophthalmic diseases, ulcers and stomach problems are other commonly reported diseases amongst the workers. It is essential to improve their living and working conditions. This awareness would generate changes in the waste handling habits of households. This will reduce the occupational health hazards providing them with better working conditions and also better economic returns. In the process their contribution towards waste management will be recognized in the society.

Singhal and Pandey (2001) focus on the various methods that could be adopted in order to make the collection process better, efficient and sustainable. They suggest targeting waste reduction at source, technological interventions and efforts towards institutional and regulatory reforms. They suggest revamping the existing collection service structure to provide community with waste bins, conveniently placed for the people to deposit domestic waste, and door to door collection of waste. This along with separation of waste, at source, into bio-degradable and non-biodegradable components would not only reduce the cost of transportation for final disposal but also provide segregated organic waste stock for waste to energy activities.

Puri Avinash et al (2008) analyse the impact of SW on community health in Jalandhar. It concludes that the SW is a major part of environment pollution; it is responsible for spreading many harmful and infectious diseases. Domestic waste, if left unattended leads to bad odour due to decomposition and spreads diseases. Waste from agriculture and industries can also cause serious health diseases because these wastes may include some chemicals, pesticides, metals etc. Uncollected waste disposed in water may contaminate it and cause water borne diseases. Plastic is also very harmful to health. The unhygienic use and disposal of plastic causes very toxic effects as some coloured plastics contain heavy metals that are very toxic e.g.: copper. The findings are that kala azar and dengue were not reported in any area whereas diseases like malaria and loose motion were reported. There are moderate health impacts on the resident population due to the SW being dumped in their surroundings. It also suggests ways in which these diseases can be prevented like proper disposal, avoiding water pollution, early diagnosis and treatment. It also emphasizes on creating public awareness in order to improve the management of SW.

Objectives

The study focuses on two main objectives i.e. collection of waste and its effects on health

1. To investigate the collection of SW in slum and residential area.
2. Analysing the effects of mismanagement SW on the health of people and children under 5 residing in that area.

We emphasize on the process of collection because there is a significant difference between the residential areas and the slum areas and as the management worsens so does the health indicators of the residents in those colonies. This issue pertains to our second objective.

Methodology

The basic source of data for the study and tool for analysis is the survey done in the areas mentioned. For our first objective we conduct a household survey on a sample of households in category A to H. This would help us to know that how and where they dispose the wastes of their house. We set the hypothesis that there exist poor collection of Solid Waste in the H category as compared to the A category and then we verify this by the household surveys. We not only capture the difference in the collection of Solid Waste in the slum area and the residential area but also the overall inefficiency of MCD to do this task. We then compare the results across categories and see if there exists a significant difference between them.

We also analyse the effect of Solid Waste on the health. We look at this dimension because SW is one of the causes for air borne and water borne diseases like tetanus, kala azar, malaria etc. This is done with the help of both household survey and doctor's survey. Two separate Questionnaires are prepared for each objective i.e. one for the households and one for the doctors. We prepare a separate questionnaire for the doctors to know which diseases are more frequent and most prevalent in the area of survey. A survey on the doctors in the private / government hospitals located near the slum area or the residential area is conducted. We can compare the results for category A and H and hence see the difference in the management and health parameters respectively. We also compare the results on health amongst children and adults.

Overview of the Study Area

In Delhi, SWM is the responsibility of three municipal bodies - the Municipal Corporation of Delhi (MCD), the New Delhi Municipal Council (NDMC) and the Delhi Cantonment Board (DCB). MCD alone manage almost 95 % of the total area of the city. The above authorities are supported by a number of other agencies. Citing and allotment of land to MCD for sanitary land filling is responsibility of the Delhi Development Authority (DDA). Solid waste utilization projects are responsibility of the Department of Non-Conventional Energy Sources (DNES) and Delhi Energy Development Agency (DEDA) under the Ministry of Environment and Forests (MoEF), Government of India. The Department of Flood Control of Delhi Administration looks after the supply of soil to be used as cover for sanitary landfills by the MCD.

We look at the area under MCD and South Delhi in particular. South Delhi presents a landscape of contrast posh colonies existing side by side with slums. Therefore, we analyze South Delhi which is further divided into wards. The colonies are further divided into 8 categories i.e. from A to H. Where residents of category A has the highest property value as compared to any other category and as we down, the value of property also goes down implying that the residents of category H have the lowest property value.)

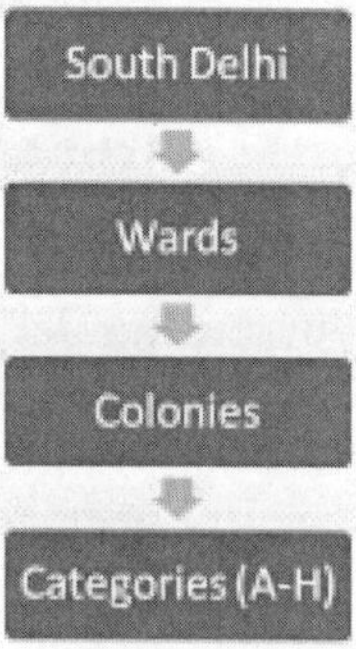

Some figures related to waste characterisation in Delhi are given below:

Table 1:Waste characterization in Delhi (2004-2005)

Compostable (%)	**54.42**
Recyclables (%)	15.52
C/N Ratio	34.87
HCV (Kcal. /Kg.)	1802
Moisture	49

Source : CPCB, 2004-05

Table 2: Area, Waste Quantity and Generation Rates in Delhi 2004-05

Population (as per 2001 census)	**10306452**
Area (sq. Km)	1483
Waste Quantity (TPD)	5922
Waste generation rate(kg/C/Day)	0.57

Source :CPCB, 2004-05

Table 3: Summary of the Study Area

Category	*Colony Name*	*Ward Name*	*Sample Size*
A	Vasant Vihar	Vasant Vihar	22
B	Sukhdev Vihar	Okhla	26
C	Sheikh Sarai	Malviya Nagar	25
C	Munarka DDA Flats	Hauz Khas	23
D	Pushp Vihar	Pushp Vihar	22
D	Satya Niketan	R K Puram	23
E	Beg Sarai	Mehrauli	23
E	Srinivas Puri	Okhla	18
F	Chirag Delhi Village	Malviya Nagar	21
F	Abu Fazal Enclave	Okhla	26
G	Batla House	Okhla	22
G	Ambedkar Nagar	Madangir	26
H	Deoli Village	Deoli	21
H	Sangam Vihar	Sangam Vihar	23

Source: Field Survey

These are also highlighted in the map of South Delhi given below:

Results and Findings

The findings are stated according to each category in comparison to the other categories. We mention the relative results. The exceptional categories under the study have been Chirag Deldhi and H (Deoli Village and Sangam Vihar). We summarize the results on the collection process in the table 4.

Table 4: CollectionProcess in the Sampled locaties.

Category	Monthly rent	Monthly charges paid	% of door to door collection	Storage facility attended daily?	Open dumping	Local community bin	Bins cleaned before overflowing?
A	>55K	100	100%	Y	N	100%	Y
B	>25K	60	100%	Y	N	100%	Y
C	9K	50	100%	N	N	100%	N
D	7.5K	50	80%	N	Y	70%	N
E	6K	50	80%	N	Y	70%	N
F	4.5K	50	70%	N	Y	N A	N
Chirag Delhi	4.5K	50	40%	N	Y	N A	N A
G	3.5K	35	65%	N	Y	N A	N A
H	2.5K	30	35%	N	Y	N A	N A

* N A = not attended at all; Source: Field Survey

1. Level of education is with each category but it is not shown in the table above. Respondents in the last category were not more than 12th pass or even illiterates in some cases.

2. It's falling consistently as we move from category A to H

3. Hand driven carts is used in the colonies of all categories to transport the waste.

4. Varieties of different colours dustbins are present only in categories A and B and in no other category. Further in categories G and H the bins are unavailable, being of different colours is out of question. Inaccessibility to community bins leads to dumping of waste on streets by households and shopkeepers residing in this area.

5. As we move from category A towards category H, the per month charges paid to kuda wala decline marginally. The lowest charge that is paid is Rs30 per month which is in sharp contrast with Rs100 per month paid in category A.

6. The households in category H and Chirag Delhi reported absence of door-to-door collection system significantly and hence, open dumping being practiced here at a large extent.
7. Open Dumping increases as we move from A to H category. There is definitely a relationship between door to door collection system and open dumping. As the door to door collection declines, open dumping increases. This trend can be observed in the table as well.
8. The problem faced due to open dumping are similar in all categories with an exception in Chirag Delhi and H category where the problem is at a critical stage as they have to accommodate for the waste from other colonies too.
9. Overflow of the waste is observed in C, D, E, F, G categories at a moderate rate but it's critically high in category H and F1 because they are not even attended by the MCD for cleaning at all.
10. No strict action is taken by MCD which is an urgent need in Chirag Delhi and the H category. But we cannot hold MCD alone responsible for it. These colonies are unauthorised and unregulated. More than an action by MCD what is required is a cooperative public action and an alternative mechanism for SWM.
11. We would also like to draw attention to the extent of problems faced due to open dumping in all categories. The problem faced in category A and B is that of foul odour only. But as we move down the line, in categories C, D, E and F, the problem is faces in sewage and presence of rodents, mosquitoes, flies, vectors etc is moderate. But the situation is worse in category G and H and at an alarming stage in Chirag Delhi and category H in particular.The problem found at the community level in category H was with the ideology of the people, their

unwillingness to pay Rs. 30 to kuda wala and disposing the waste in open dumping. This further encourages the action of open dumping amongst other people as well.

Now we compare the results obtained on the health indicators with the help of the household survey. We make this comparison only for categories through C to H. We first make comparisons for the children under the age of 5. We, then compare the occurrence of fever, cough and cold in a year on an average.

Table 5: Locational Pattern of diseases

Diseases	C	D	E	F	Chirag Delhi	G	H
Diarrhoea	20	20	22	20	35	25	35
Malaria	5	10	10	15	20	15	20
Dengue	5	10	10	15	20	15	20
Dysentery	10	10	15	20	30	20	25
TB	-	-	-	-	-	-	5 - 10
Asthma	10	10	15	20	30	20	20

Source: Field Survey

Other prevalent disease in children under 5 are asthma, chickenpox, diarrhoea, pneumonia, tetanus, typhoid, malaria and measles. In category C, the most prevalent disease is malaria, in category D, persistent fever is followed by asthma, chickenpox, diarrhoea and pneumonia will similar prevalence rate. In category E, results are similar to observation made in category D. In category F in particular; diarrhoea is the most prevalent disease followed by persistent fever, asthma, chickenpox and pneumonia. Residents also suffer from bronchitis and influenza in this category. Along with that, the difference in the prevalence rate is relatively much higher than what is observed in the other categories, which implies people fall ill more frequently here than in other categories. The situation in category G and H is even worse but better than Chirag Delhi. Here also the most prevalent disease is diarrhoea followed by bronchitis and pneumonia. Prevalence rate in percentage are as follows.

Table 6: Locational Pattern of diseases (in Percentage)

Diseases	C	D	E	F	Chirag Delhi	G	H
Asthma	5	5	10	15	40	20	35
Chickenpox	5	5	10	20	40	25	40
whooping cough	10	10	20	25	40	20	35
Diarrhoea	10	10	20	30	70	35	45
Pneumonia	10	10	20	25	40	25	30
Typhoid	5	5	10	25	30	20	25
Malaria	5	10	10	15	40	25	25
persistent fever	20	20	30	40	60	40	50
Measles	10	10	10	30	40	30	35
Influenza	-	20	20	35	40	35	25
Bronchitis	20	20	25	30	50	40	40

Source: Field Survey

Table 7: Prevalence Rate of diseases for all member of the households

Category	fever	cough	Cold
C	2- 4	2 – 4	2 – 3
D	3 – 5	2 – 3	2 – 3
E	3 – 5	3 – 4	2 – 3
F	3 – 5	3 – 4	2 - 3
Chirag Delhi	>10	>5	>5
G	4- 6	3 – 4	2 -3
H	>5	>5	>5

Source: Field Survey

In category C, the disease with highest prevalence rate is dysentery, diarrhoea and bronchitis. In category D too it is diarrhoea, bronchitis, malaria and dengue. E category follows results similar to D category. In Chirag Delhi we observe a high prevalence rate of diarrhoea, dysentery, bronchitis and malaria. As observed earlier, the situation in category G and H are worse but better than Chirag Delhi.

Table 8: Prevalence Rate of diseases for all the embers of the household

Category	fever	cough	cold
C	5 - 6	4 - 5	3 - 4
D	8 - 10	>5	>5
E	9 - 10	>5	>5
F	9 - 10	>5	>5
Chirag Delhi	>20	>10	>10
G	10 - 20	>5	>5
H	>15	>5	>5

Source: Field Survey

These results are consistent with the findings in the process of collection. As we have observed that the problems faced due to open dumping are severe in Chirag Delhi and H category which in turn gives way to rodents, insects, flies, mosquitoes and other bacteria carrying vectors. The occurrence of fever, cough and cold is highest in these categories. One of the respondents in the category H said "yahan koi nahi ata kuda saaf karne ke liye aur machar saare time rehte hai".

Similar results were found in the doctor's survey too. The table below concludes the findings in categories E, F, G, and H

Table 9: Results from Doctors' Survey

	E	F	G	H
Max no. of diseases seen	Diarrhoea Fever Cold	Diarrhoea Fevr Cold Malaria	Darrhoea Fever Cold Malaria Influenza	Diarrhoea Fever Cold Malaria Influenza Dengue Leprosy Asthma

Contd...

Max no. of diseases seen (children<5)-	Diarrhoea Persistent fever	Diarrhoea Persistent fever	Diarrhoea Persistent fever Phenumoni Whooping cough	Diarrhoea Persistnt fever Pheumonia Whooing cough Malaria
Reason for the disease (Polluion, hygiene and unsafe drinking water)-	Yes	Yes	Yes	Yes
Disease with which patients revisited	Persistent fever	Persistent fever	Persistent fever Malaria	Dengue Asthma Persistent fever
Suggestions	Removal of open dumping and safe drinking water	Removal open dumping and safe drinking water	Removal of of open dumping and safe drinking water	Removal of open dumping and safe drinking water

Results are similar to the household survey

Sugestions and Policy Recommendations

1. Steps should be taken to install different coloured dustbins (or at least a community bin in the bottom most categories) where ever required. To accomplish this task, the government should join hands with NGO's. The bins should be located at a central point to make it accessible to maximum number of people.
2. Collection facilities should be improved, open dumping should be discouraged.
3. If the need of an open dumping is inevitable in a particular area, then it should be strategically chosen so that it has minimum effect on the health of the residents and yet .The area chosen for open dumping should not be near to any place of public gathering

like temples, market, milk booth etc. and it must be the case that it's visited only for the disposal of waste and no other purpose. Special attention should be paid towards the fact that children do not play around such areas.

4. Public awareness should be created about the importance of hygiene and cleanliness. For example, practice of covering the bins can help to reduce the possibilities of many diseases.

5. Special attention should be paid to the health factor in the slum areas. This can be done by regular visits by the doctors in the nearest government hospital.

6. Free vaccines should provided for the preventable diseases like tetanus, hepatitis etc.

7. A market mechanism should be created where the people have an incentive for proper disposal of the waste. For example, there exist an unorganised markets of kabadi wala, where the households sell old newspapers and plastic bottles etc. for a nominal amount. Even though however small, any monetary incentive given to the households can help in reducing mismanagement of waste.

8. Compulsory education at all level may be adopted in order to protect environment for the waste pollution.

Conclusions

- Different coloured bins are not located in category C, D, E. No bins are available in F, G, and H category at all; different coloured bins are out of question.

- Mismanagement of SW increases as we move down the category. The situation is worse in Chirag Delhi and demands immediate attention. Open dumping is practiced at a large extent by the residents here due to absence of door to door collection.

- The trend observed in the survey suggests high negative relationship between door to door collection and open dumping. Conditions are similar in Sangam Vihar and Deoli Village.

- Urgent cleaning of local collection points and open dumping is needed in Chirag Delhi, Sangam Vihar and Deoli Village. These areas are not visited even once in a month by MCD and the dumping and collection point generally overflows and waste litters around in the surroundings.
- The ignorance of the municipal authority in these three areas has lead to severe impact on the heath of the residents here. Frequency of fever, cough and cold in Chirag Delhi, Sangam Vihar and Deoli Village is as high as 20 as compared to that of 5 in category C and D.
- The most prevalent disease is found to be diarrhoea. Its frequency is highest in the monsoon season, which can be related to SW. It is one of the reasons for the disease, during rains, the bacteria transmits faster and hence spreads diseases.
- M&H Rules 2000, it has been a decade that it has come into effect, but it has very weak implementation in all categories except A and B. The reason behind good indicators here is due to the efforts of the resident's welfare association.

References

Agarwal Ankit, Ashish Singhmar, Mukul Kulshrestha, Atul K. Mittal (2005), "Municipal Solid Waste Recycling and Associated Markets in Delhi, India" , Science Direct ,*Rresource Conservation and Recycling*,Vol 44,Issue 1 PP 73-90.

Ali Mansoor and Marielle Snel(1999), "Lessons from Community Based Initiatives in Solid Waste", WELL London School of Hygiene & Tropical Medicine, UK WEDC, Loughborough University,March

CPCB (Central Pollution Control Board) (2004-05), "Status of Solid Waste Generation, Collection, Treatment and Disposal in Metro-cities" CPCB, New Delhi

Delhi Planning Board "Delhi Urban Environment and Infrastructure Improvement Project", Delhi Planning Board, New Delhi, chapter 10.

Ministry of Urban Development "MSW Rules, Report of Technology Advisory Group on Solid Waste Management", Constituted by the Ministry of Urban Development, Government of India.

Papiya Sarkar,(2003), "Solid Waste Management in Delhi - a Social Vulnerability Study", Toxics Link, New Delhi, India.

Puri Avinash , Manoj Kumar, Eonkar Johal (2008), Solid Waste Management in Jalandhar City and its Impact on the Community Health, IJOEM (Indian Journal of Occupational and Environnmental Medicine, Vol 12, Issue 2.

Ramchandra T.V and Shruti Bachamanda (2007), "Environmental Audit of MSWM," *International. Journal of .Environmental Technology and Management, Vol. 7, Nos. 3/4, 2007.*

Sharholy Mufeed , Kafeel Ahmad, Gauhar Mahmood , R.C. Trivedi (2008), "'Municipal Solid Waste Management in Indian Cities - a Review".Science Direct , waste management, Vol 28, issue 7, pp 1276-1287.

Singhal Shaleen and Sunil Pandey(2001), "Solid Waste Management in India: Status and Future Directions", TERI, New Delhi.

Vishwanathan C and J. Tränkler (2003), "Municipal Solid Waste Management in Asia - A Comparative Analysis", workshop on Sustainable Landill Management 3-5 December, 2003, Chennai, India , pp 3 -15.

Chapter 13

An Analytical Overview of Solid Waste Management: A Case Study of Delhi

Subhash Anand

Introduction

One of the important aspects of urbanization in India is the tendency towards concentration of population in larger urban centers. It leaves several issues unmanaged concerning mainly with water supply, solid waste management and environmental decay etc., which get manifested in complex web of social problems and dilemmas of urban development. The level of infrastructure and services in urban centers has lagged behind the pace of urbanization. There is deterioration of the quality and standard of urban services. Solid waste management is an obligatory function of urban local bodies in India. Infrastructure development is not in a position to keep pace with population growth and requirements leading to low level of waste collection efficiency. Lack of financial resources, organizational weakness, improper choice of technology and public apathy towards waste management have made this service far from satisfactory. In future, as the economy improves, the increase in quantity of garbage due to population increase will be reinforced by the rise of per capita income. Additionally, with relative prosperity, more of non-bio-degradable waste could be produced in future.

Waste problems in cities include the increasing difficulty of acquiring new land areas for disposal, the generation of pollution from processing and disposal of waste, disposal caused resources depletion and the huge cost involved in waste processing. Uncontrolled dumping of wastes on outskirts of towns and cities has created overflowing landfills, which are not only impossible to reclaim because of haphazard manner of dumping, but also have serious environmental implications in terms of water contamination, land degradation and air pollution contributing to global warming. Environmental degradation is taking place and organizations that are responsible for environmental management are facing many problems and challenges. Certain environmental parameters fast becoming critical for sustainable development of urban areas.

A pertinent question is to what extent the public sector is capable for handling the situation of increased pressure on natural resources and environmental degradation. Due to weak public sector performance, the management of some public services might be transferred to other entities, with better accountability and responsibility. It is often argued that entities located in civil society are more responsive to people's need and therefore they can deliver the required services more efficiently than the public sector. Now, governments have begun to look for public-private partnership arrangements with people and non-governmental organizations (NGOs) participation in order to bring sustainable development of urban environment.

Nature of Problem

As of today, solid waste has emerged as a major environmental concern from local level to global level. At the household level, people discard off their waste in an improper manner. The methods of collection, segregation, transportation and disposal of wastes are not up to the mark and have resulted in creating an unhygienic conditions in surrounding environment. A number of Indian cities adopt the practice of dumping of waste in open landfill sites. In Delhi, sixteen landfill sites have filled up already and existing sites (Bhalswa, Gazipur and Okhla) have reached to saturation point and overflowing. Acquisition of new sites is a very tough task for the authorities. Out of total municipal solid waste of Delhi, about 95 per cent of waste dispose off at the landfill sites, remaining only 5 per cent go to the processing. It is a common sight to find out domestic and non-domestic solid wastes dumped in haphazard manner.

There are potential risks for human health and environment from inefficient collection and unscientific disposal of municipal solid wastes. Direct health risk expose mainly to the sanitary workers and rag pickers- an essential link in the urban solid waste collection and recycling process. By working in the most unhygienic under vulnerable environment, rag pickers arc really providing a great unthankful service to the society at the cost of their life. Besides, social marginalization, waste workers and their families are subject to economic insecurity, occupational health hazards, lack of access to normal social services etc. For the general public, the main risks to health are indirect and arise from the breeding of disease vectors. More serious, is the transfer of pollution to water, air and land resources.

The rate of generation of municipal waste has increased a lot in spatio-temporal context that urban local bodies are unable to deal with the tremendous amount of garbage generated in Delhi. As a result, a significant amount about 15-25 per cent of garbage, remains uncollected and accumulates in the form of heaps at various places. It soon begins to rot and becomes an environmental and health hazard. The municipal solid wastes generated in Delhi are about 7000-9000 metric tonnes per day and quantity of this has been consistently rising over the years. With the increasing population, urbanisation, rising socio-economic status of people, changing life styles are attaining larger and complex dimensions, requiring integrated and sophisticated interventions in terms of technologies, policies, plannings, rules, regulations and legislations. The composition of municipal solid waste has changed in such a manner that, today, a major proportion of the waste is composed of non-biodegradable materials. The present system of recycling of plastic, glass, metal, paper etc. is largely informal. This category of waste can be recycled or disposed off through special processes. Due to these reasons, the task of handling solid waste has become a highly specialized managerial task. The waste created by a person may be resource for another one. The most important problem is that, it is seldom taken as a resource, it is just taken as a pollutantwhich is to be dumbed,Therefore, it depends upon human beings, whether it should be considered as a resouruce or a resistance.

Sustainable solid waste management requires not only efficient collection, proper disposal but also waste reduction, reuse, recycle and resource recovery. The important aspects which require priority are to improve the management and organizational capability and enhance people's participation and the handling of waste in an appropriate and scientific manners. Now, it has become critical issue of Delhi, which requires high level planning and management. The overall goal of waste management programmes must be welfare and safeguard of public health, waste minimization, maximization of resource recovery and protection of environmental quality. Therefore, it is necessary to conceive the integrated solid waste management plan along with urbanization process rather than tackling it after the problems appears.

Study Area

The area selected for the present research study has a unique position with regard to its location. Physiographically, two important features of Delhi are river Yamuna and the extension of Aravalli hills. Being the land locked area, it is characterized by extreme dryness with hot summer and cold winter. It is surrounded by Ghaziabad, Sonepat, Faridabad, Gurgaon, Rohtak and Ballabgarh. Delhi spread over an area of 1483 square kilometers (0.04 per cent of total geographical area of India) and out of this Municipal Corporation of Delhi (MCD) occupies an area of 1397.29 sq.km. (94.22 per cent), whereas New Delhi

Municipal Council (NDMC) and Delhi Cantonment Board (DCB) possesses 42.74 sq.km. (2.88 per cent) and 42.97 sq.km. (2.90 per cent) respectively. Delhi's 96.96 per cent population lives in the area under jurisdiction of MCD, whereas 2.14 per cent and 0.99 per cent population lives in NDMC and Delhi Cantonment area respectively. MCD is among the largest municipal bodies in the world. It is next only to Tokyo in terms of area. MCD has the dual responsibility of providing civic services to its people in both rural and urban areas, resettlement colonies, regularized and unauthorized colonies, JJ squatter etc. MCD has divided its entire area into 12 zones, namely, City, Central, Civil Lines, Karol Bagh, Narela, Najafgarh, Rohini, Shahdara (North), Shahdara (South), Sadar Paharganj, South and West zone. Geographically, Najafgarh zone is largest followed by Narela, Central zone " whereas Sadar Paharganj and City zones are smallest. These MCD zones are different from administrative districts and subdivisions made by Government of India. Each zone is further divided into a number of wards. Out of the total area under MCD, 57.09 per cent is characterized as rural and remaining is as urban area. NDMC has bifurcated its area into two circles, North and South. Delhi Cantonment area is also divided into civil and military area. Out of total area of Delhi, 924.68 sq.km. is classified as urban and 558.32 sq.km. as rural area.

According to 2001 Census, Delhi had 13.85 million population with a growth rate of 47 per cent. Of this, 54.95 per cent were males and remaining were females. 1,29,05,780 were urban dwellers constituting 93.01 per cent and remaining 9,44,727 persons were living in rural areas. Delhi has undergone a rapid phase of urbanization and some parts of the city are now densely populated. Among all districts of Delhi, New Delhi and Central district were fully urbanized followed by East and West district. North-West and South-West were lesser urbanized. Maximum density of population was reported in MCD (urban) area whereas NDMC and Delhi Cantonment occupied second and third position respectively. Total 25,54,149 households were recorded of which 93.4 per cent were urban and 6.6 per cent were rural households.

There are mainly three local bodies (MCD, NDMC and DCB) responsible for management of municipal solid waste in their respective jurisdiction. Around 0.1 million rag pickers including women and children are engaged in waste picking and saving the big amount of money of concerned authorities. Various agents of both sectors public and private are active. The Conservancy and Sanitation Engineering (CSE) Department which is also known as Department of Environmental Management Services (DEMS) has employed about fifty thousand sanitary workers for collection and disposal of solid waste in its area. Besides, a large staff both skilled, semi-skilled mid unskilled tu supervise. A large number of infrastructure associated with garbage management arc available. At present only three composting plants (Bhalswa, Okhla and Tikri Kliurd) are there which are running below their capacities.

Aims and Objectives

The main objectives are as follows:

(1) To study the spatial and temporal aspects of waste generation and collection.

(2) To analyse the nature and magnitude of municipal solid waste.

(3) To know people's perception regarding solid waste management.

Hypotheses

(1) Quantity of waste generation starts decreasing with increasing distance from urban area.

(2) Lack of awareness results to low level of public participation in sustainable waste management.

Data Base and Research Methodology

Keeping in mind the aims and objectives of this research study, it is based on primary as well as secondary data sources. The primary data have been generated through structured questionnaires and field observations. Two field surveys have been conducted, one for household and second for rag pickers. For the households survey, out of total twelve zones of Municipal Corporation of Delhi, New Delhi Municipal Council area and Delhi Cantonment area, six zone have been selected on random basis. These selected zones were South, Shahdara (South), Rohini, Poschim Vihar, Civil Lines and Shahdra (North) of MCD. A cross sectional primary data have been created from higher income group (HIG) , Middle income group(MIG) , and lower income group(LIG),families. In order to find out different income group localities, criteria of Delhi Development Authority (DDA)HIG flats and MIG flats has been adopted for HIG and MIG families respectively. Resettlement colonies have been taken into account for LIG household survey. The HIG households have been selected in Vasant Kunj of South zone and Mayur Vihar phase-Ill of Shahdara (South) zone. Rohini sector XVIII of Rohini zone and Paschim Vihar of West zone chosen for MIG household survey. While, Jhangir Puri resettlement colony of Civil Lines zone and Nand Nagri of Shahdara (North) zone of MCD have been selected for the LIG primary survey in study area. 50 households from each locality and total 300 households have been surveyed randomly from six zones of MCD, including 100 each families from HIG, MIG and LIG localities. The respondents have been interviewed through household questionnaire, which was designed to gather a broad range of informations regarding solid waste management in their respective localities.

It is matter of fact that rag pickers are playing a vital role in the management of solid waste in Delhi from economic and environmental point of view. Therefore, another interview schedule was formulated for the rag pickers. Second primary survey was made among rag pickers engaged in picking of waste recyclable materials in study area. Total 100 rag pickers have been surveyed from various locations, viz. landfill sites, street, waste receptacles. During the both surveys, one hand information were collected through direct personal contact.

In order to fulfill the aims and objectives of the present investigation, the relevant and useful data informations have been obtained from secondary sources both in published and unpublished form. Many organizations were visited and interviewed. The background informations pertaining to extent, area, demographic characteristics, urbanization and socio-economic profile have been taken from

District Census Handbook, Delhi Statistical Handbook, Census of India. Required data and comprehensive information and published available materials related to solid waste management in spatio-temporal context have been collected from Municipal Corporation of Delhi (MCD), Central Pollution Control Board (CPCB), New Delhi Municipal Council (NDMC), various departments of Government pf National Capital Territory of Delhi (GNCTD) and Centre for Science and Environment (CSE). Apart from these organizations, other informations were also gathered from National Environmental Engineering Research Institute (NEERI), National Institute of Urban Affairs (NIUA), The Energy and Resources Institute (TERI), Infrastructural Leasing and Financial Services (IF&FS). Data and informations also collected from non-governmental organizations such as Asian Centre for Organization Research and Development (ACORD), Indian Society of Environmental Management (ISEM), Indian Environmental Society (IES), Vatavaran, Shrishti, Chinton etc. A large number of published reports, journals and books were also consulted. After completing the field surveys, the processing of data was taken up with various suitable statistical techniques. The Waste Service Index (WSI) has been prepared based on Z-score methods on the basis of services provided by MCD in various zones and best served, moderately served and least served zones were identified.

Results and Discussion

Municipal solid waste management is a complex system of various interrelated activities involving different actors i.e. government, people, rag pickers, and non-governmental organizations. Composition and quantity of municipal waste generation depends upon food habits, life style, population size, level of urbanization, land use, socio-economic status etc. Actual waste generated per capita is higher as significant amount of garbage lying uncollected and picked up by rag pickers. Among the all local bodies, NDMC has highest production of garbage in terms of per capita per day. All the high population density zones (City, Karol Bagh, Sadar Paharganj) are having high rate of municipal waste production. Municipal Solid Waste of Delhi is highly bio-degradable in nature which has higher moisture content and low calorific value and it is more suitable for composting purposes. Fifty Nine per cent growth in amount or waste generated has been recorded during last 11 years but infrastructure has not increased proportionately. Accuracy increased in data of waste generation after introduction of weighbridge machines by MCD in 2000 at all three landfill sites.

There is no clear cut standard for the placement of waste receptacles. Installation of dustbin is seen as arbitrary. People were not using dustbins as per their purpose. Present design and location of waste receptacles is not accepted to the people. The spatial analysis of the number of dustbins, garbage carrying vehicles and sanitary workers reveals that their number gradually declines from core to periphery zones.

According to Waste Service Index based on Z-score value, Karol Bagh, Sadar Paharganj and Shahdara (S) are identified Highly Served Zones whereas Najafgarh and Narela are Least Served Zones. Remaining all seven MCD zones are categorized Moderately Served Zones. Najafgarh and Narela zones have more potentialities to enhance the infrastructural facilities related to municipal solid waste disposal.

Ninty to Ninty five per cent waste of Delhi is reaching landfill sites and increasing the height of sites. Presently, all three landfills are fully packed and overflowing. Among all three present landfill sites, Bhalswa is receiving maximum (44 per cent) garbage from six zones of MCD whereas Gazipur and Okhla are getting 37 and 19 per cent garbage of Delhi respectively.

Disposal of household waste in government dustbins was found a common practice. In study area, private garbage collectors were dominating. Somewhere, due to absence of dustbins people disposing their waste wherever they find space, which in due course of time turns into collection point.

Status of solid waste disposal was comparatively better in higher income group localities but situation was critical in lower income group areas. People of higher and middle income group localities were highly conscious about their health and surrounding environment. They see uncollected garbage as a potential health problem. Majority of waste collection points were observed with stray animals (cows. doges, pigs) and rag pickers. Open sites and open site and may *dhalaos and* dustbins in MCD areas were found totally unhygienic and unsafe for environment and human health. Sometimes stray animals cause high danger to people particularly rag pickers and sanitary workers. Hot spots were found in critical situation. Seventy one per cent people considered , Solid Waste Management is a critical problem in Delhi, which needs immediate action.

Despite of numerous legal initiatives taken by government to improve the situation, desired success has not taken place far. Government launched many campaigns time to time but failed because of lack of people's participation. Level of awareness among people was found low. But people were found interested to participate in solid waste management streams.

Fourty six per cent of people were not satisfied by providing MCD services and maximum of them were from LIG and MIG colonies. More than 50 per cent people admitted that Right to Information (RTI) Act will be proved as a weapon for making pressure on local bodies in order to make the improvement in collection and disposal of garbage.

About fourty per cent people suggested that awareness and participation among people must be enhanced. Ninteen per cent suggested that accountability and responsibility of MCD should be fixed. Only 11 per cent wanted that MCD should be privatized.

Rag pickers were working in extremely unhygienic environment. They have lower socio-economic conditions. They are playing vital role in solid waste management sector. They have been exploited by junk dealer, police and MCD personnel. They do not have any formal organization or support from government. Fourty seven per cent rag pickers were suffering from various diseases. Fourty four per cent were not knowing that they are exposed to dangerous germs waste picking About Ninty per cent wanted to change their occupation.

Conclusion

The composition and magnitude of municipal solid waste depends on a wide range of both external and internal factors such as as geographical aspects, standard of living, food habits, cultural traditions, socio-economic status etc. Solid waste management is essential task which has important consequences for public health and well being, the quality and sustainability of the urban environment and the efficiency and productivity of the urban economy. Health and environment degradation and growing awareness of the public is putting more and more presence on municipal bodies to find sustainable solutions to solid waste management problems. If considerations were given from the grass root level, then problems would be solved to the large extent. Today, the need for the integrated solid waste management strategy is being felt at all levels. This strategy involves reduce, reuse, recycle, resources recovery, processing and waste to energy. Segregation of municipal waste at source should be mandatory for all and chain of segregation should not be broken from household to landfill sites.

References

Abd Qdais, H.A. et al. (2001), "Analysis of Residential Solid Waste at Generation Sites" in Jane C. Powell et al. (eds.). *Waste Management and Planning,* Elgar Publishing Limited, U.K.

Agrawal, Ankit et al, (2005), "Municipal Solid Waste Recycling and Associated Markets in Delhi, India". *Resources Conservation and Recycling,* Vol. 44.

Ahmed, S.U. (1980). "Urban Problems and Government Policies" A Case Study of the City of Dacca" in Ballhatchet, K, and Harrison, J, (eds), *The City in South Asia,* Premodern and Modern. London.

Ali, Sabir (1995), *Environment and Resettlement Colonies in Delhi,* Council for Social Development, New Delhi.

Anand, Subhash (2005). "Solid Waste Generation and Management in Delhi; A

Sustainable Approach" in Singh, J. (ed.), *Environment and Development,* I.K. International Publisher, New Delhi, pp. 621-639.

Ansari, J.H. (1999)' "*Solid Waste Management in Delhi - Need for Partnership Arrangements*", School of Planning and Architecture, New Delhi.

Asian Age (2003), "Delhi to Get First scientific Sanitary Landfill Site". *The Asian Age,* Daily Newspaper. 13 August. Delhi.

Baht,R.V. and Moy, G.G. (1997), "Monitoring and Assessment of Dietery Exposure to Chemical Contaminants", World Helath Statistic, Quarterly, 50 (1-2): pp 132-149

Bansal S.P. (1998), "Planning for Solid Waste Management in Delhi"? *Spatio-Economic Development RecordDevelopment Recod,* Vol .5, No.3 , May-June

Basu, Swapna (1998). "Solid Waste Disposal and Sanitation in Durgapur" Geographical Review of India, Vol.62, No. 1 *Geographical Review of India.* Vol. 62, No.1

Beukering, Pieter Van (1994), "The Recycling sector in Bangalore; An Economic Analysis of Different Types of Formal and informal Enterprenuers, Recovering Urban Solid Waste" in Baud,I. and Sachenk, H (eds)m *Solid Waste Management,* Manohar Publishers , New Delhi, pp 108

Bhattacharya, S.K.(1997), "Management of Solid Waste",Yojana, Vol . 41, No. 9

Bhide,, A.D. (1994). "Solid Waste Management in Urban Areas". *Nagartok,* Vol. XXVI,Oct-Dec.,No.4.

Central Pollution Control Board (2006). "*Assessment Status of Municipal Waste Management in Metro Cities and State Capitals*", New Delhi.

Chaturvedi, Bharti (2007), "Go Dutch to Manage Waste", *The Hindustan Times Daily Newspaper, 29 Jan, Delhi*

Cointreau, *S.J. (1987), "Living with Garbge: Cities Learn to Recyle" Development Forum.* Jan-Feb. pp. 12-13.

Cointreau, S.J. (1987). *Solid Waste Recycling: Case Studies in Developing Countries,* Washington, D.C.

Dankelman, I. and Davidson, J, (1998). *Women and Enviroment in the third Women and Environment in the Third World: Alliance for the Future,Earthscan Publication Ltd.,London*

Dhamija, Urvashi (2006), "*Sustainable Solid Waste Management: Issues, Policies and Structures*", Academic Foundation, Delhi.

Ghosh, Archana (2000), "*Solid Waste Management in Delhi: An Exploratory Study a Local Government - Community Interface*", Indian Social Institute, New Delhi.

Gotosh, S. (1989), "Issues and Factors to be Considered for Improvement of Solid Waste Management in Asian Metropolies", *Regional Development Dialogue,* 10(3).

Harris, F. (2004) (eds.), *"Global Environmental Issues"*, John Wiley & SonsPublishers, England. • ,

Indian Environmental Society(IES) (2001), *"Solid Waste Management: A TrainingManual"*, IES, Delhi.

Indian Society of Environmental Management (ISEM) (2000). *"Development of Statistics in Environment Sector: Solid Wastes"*, ISEM, New Delhi, pp. 175.

Jain, A.K. (2000). "Solid Waste in Delhi, MCD, NDMC", *Civil Affairs,* Vol. 31, No. 12.

Kaushal, A. et al. (1998), "Delhi Municipal Solid Waste and Environment - An Appraisal", *Indian Journal of Environmental Protection,* Vol. 18, No. 2.

Khan, M.U. (2004), "Regulations and Management of Municipal Waste: Indian Scenario", *NIST ADS News,* Vol. 6, No. 1, p. 43.

Krishna, D. and Sangeetha, T.L. (1994). Approaches to Urban Solid Waste Management: Linkages between Formal and Informal Systems of Source Separation and Recycling - A Profile of Hyderabad, Discussion Paper for the Workshop on "Linkages in Urban Solid Waste Management", Bangalore, pp. 39-41.

Ministry of Environment and Forest (2000), *"Municipal Solid Waste Management Handling Rules"*, Ministry of Environmental and Forest Notification, New Delhi.

Municipal Corporation of Delhi (2004). *"Bhagidari Workshop on Municipal Solid Waste Segregation"*, Report 8-10 Jan., MCD, Delhi, pp. 13-15.

National Environmental Engineering Research Institute (1996). *"Solid Waste Management in MCD Area"*, Phase-I Report, NEERI, Nagpur.

Sarkar, Papiya (2003), "Solid Waste Management in Delhi - A Social Vulnerability Study" in Martin, J. Brunch et al. (eds.), *Proceedings of the Third International Conference on Environment and Health,* Chennai, 15-17 Dec, Department of Geography, University of Madras and Faculty of Environmental Studies, New York, pp. 451-464.

Subhash (2007), *"Generation and Management of Municipal Solid Waste in Urban Delhi: A Geographical Analysis"*, Ph.D. Thesis submitted in Department of Geography, University of Delhi, Delhi.

The Times of India (2006). "Waste Management in a Complete Mess", *The Times of India,* Daily Newspaper, May. 18, Delhi.

US Environmental Protection Agency (2004), "*Municipal Solid Waste Facts*,USEPA,accessedat http://www.epa.gov/epao/swer/nonhw/muncpl/facts. htm.

Venkatesan, R. (2006). "A Menance on Islands - Disposing Plastics", *Science Reporter*, Vol. 43, No. 6, June, pp. 34-36.

Chapter 14

Sustainable Decentralized Solid Waste Management

Mohammad Tariq

Introduction

What we discard, reject, abandon or otherwise release into the environment becomes waste. We usually call this garbage. If not managed properly, these waste materials can have adverse impact on the environment and serious consequences on the public health, arising from the contamination of the soil, water or air and spread of diseases through vectors thriving on waste. The problem of urban solid waste is assuming alarming proportion. Municipal solid waste consists of households waste, construction and demolition debris, sanitation residue and waste from streets. This garbage is generated mainly from the residential and commercial complexes.

Situation of waste in India

Management of Solid Waste (MSW) is a serious challenge being faced by the urban and rural communities all over the world. It has acquired grave proportions in urban areas, particularly in the large urban settlements. In 1947, cities and towns in India generated an estimated 6 million tones of solid waste; according to various estimates in 1997 it is about 48 millions tones, which is likely to grow two-folds in another 15 years or so. As per Tata Energy Research Institute (TERI) estimates daily per capita waste generation in 1995 stood at 0.45 kg, growing at the rate of 1.33 per cent per capita/year. The average per-capita waste generation varied across various class-sizes of urban areas, whereby, the larger one's generated higher quantities as compared to the smaller towns. Similarly, the urban poor generate nearly one-fourth less waste as compared to the affluent.

With the growing urbanization and change in life style and food habits, the amount of municipal solid waste generated has been increasing rapidly, and its composition has been continuously changing with population redistribution, changing lifestyles, income, and consumption patterns in Indian towns and cities. While the share of paper, plastics, rubber, glass and metals besides several multi-layered materials is constantly increasing, that of the bio-degradable material still remains significant in our towns and cities. More than 25 per cent of the municipal solid waste is not collected at all, 70 per cent of the Indian cities are lacking in the capacity to transport it, and there are no sanitary landfills to dispose of the waste. Existing landfills are neither well equipped or managed nor lined properly to protect against contamination of soil and groundwater.

Existing Policy

Ministry of Environment and forests Government of India through the Gazette notification of 25th September 2000 has brought into force the Municipal Solid Waste (Management & Handling) Rules, 2000, applicable to every municipal authority responsible for collection, segregation, storage, transportation, processing and disposal of municipal solid waste. The Municipal Solid Waste (Management and Handling) Rules 2000 makes it mandatory for every municipality to provide door-to-door collection of solid waste and segregation of bio-degradable waste from dry waste and other rejects. The organic (wet) waste shall be processed through the biological composting and the inorganic waste (dry waste) shall be process through recycling. Land filling should be restricted to waste that is not suitable either for composting or recycling.

In spite of several policy and organisational changes pertaining to the role of the local bodies, particularly the municipalities, collection, storage, transportation and disposal of solid waste remains one of the major problems in urban India. The traditional method of collection and disposal in landfill sites is neither appropriate nor feasible in the contemporary context. The land required for landfills has already shrunk in view of the changing land-use patterns besides innumerable, hazardous and insurmountable environmental and health problems.

Method of Waste Management

The objective of Municipal Solid Waste Management is to reduce the quantity of solid waste disposed of on land, by recovery of materials and energy from solid waste in a cost effective and environment-friendly manner. The solid waste management practices have continued to remain inefficient and outdated. There are two approaches on which a waste-management strategy can be premised: Centralised and Decentralised.

Centralised and Decentralised Waste Management

In India, the management of solid waste has largely been the responsibility of the local governments, i.e. municipalities. The arrangements made by the municipalities have remained grossly inadequate and inefficient and largely confined to collection and disposal of waste. Studies have indicated that a significantly large proportion (60-70 per cent) of waste generated in our cities are seldom collected and disposed off leaving our streets, roads and neighbourhood littered with stinking waste attracting animals and birds, thus aggravating the intensity of the problem. Moreover, our municipalities are cash starved besides being organisationally incompatible to meet the current and future challenges emerging from the burgeoning solid waste scenario in India. One may also note that expensive technologies for the treatment of waste in the context of India, besides having several environmental and social implications have not been found to be economically viable.

Decentralization pertains to the process of devolution of resources and decision-making powers through planned intervention at the local level defined on the basis of a geographical boundary. Residents of such well defined territory form the local community who democratically participate, take decisions, monitor and share the cost fully or partially for certain activities and functions which can be performed more efficiently at the local level. Decentralised Solid Waste Management Systems (DSWMS) needs to be appreciated as a strategy, which aims at providing solutions to a large number of problems and issues mentioned above. Decentralized solid waste management strategy endeavours to bring together various stakeholders such as the municipalities, local community, waste collectors and Non Government Organisations (NGO)/CBO (Community Based Organisations) for achieving sustainable, environment friendly and economically viable solutions whereby the citizens gradually begin to look at wastes as a resource rather than refuse. It can if carefully implemented, can change the perception of the people about waste. There is need for paradigm shift in our approach to waste management from highly centralised to decentralised strategies.

Zero Waste Management strategy is geared towards bringing about this model change in the traditional waste management approach. Zero waste management combines ethical practice with a solid economic vision, both for local communities and other stakeholders. On one hand, it creates sources of livelihood and businesses, which collect and process secondary materials into new products and on the other, it offers major corporations a way of increasing their efficiency, thereby reducing their demands on virgin materials as well as their waste disposal costs. It aims to reduce, reuse and recycle solid wastes and change perceptions about waste. Several such initiatives have been successfully accomplished in various parts of the world, which shows ways to reduce transport costs, improve collection services and provide higher and regular income and better working condition to waste pickers, besides helping in keeping the urban environment clean and healthy. The waste pickers who have traditionally been actively involved in informal collection and recycling of solid waste either from the landfill sites, streets and municipal bins need to be formally incorporated into municipal solid waste management systems. Decentralized and integrated solid waste management systems are likely to reduce the burden of the municipality and enabling the local bodies move towards organised waste collection.

Segregation - Key for Managing Waste

Waste is not a homogenous entity. However, we can broadly classify it as wet waste (biodegradable) and dry waste (non-biodegradable).

Wet waste -The decomposition of organic matter by biological action has been taking place in nature since life first appeared on our planet. Materials such as vegetable matter, animal manure and other organic refuse, otherwise wasted materials, can be converted into a more stable form for use as a soil amendment through the process of composting. The final product of the composting process is called compost. Generally speaking, there are two kinds of decomposition of material that yield compost - aerobic and anaerobic. Composting facilities are relatively cheap and easy to operate. They can reduce the total amount of municipal solid waste by 40 per cent, thus reducing the amount of garbage to be disposed of in landfills. Composting can be carried out individually or commercially in units that may be centralized or decentralized. There are many techniques that are used for composting.

Dry waste - Dry waste can be either recyclable or non-recyclable. Recyclable waste can be sold in the market and non-recyclable waste has to be sent to a landfill. Communities are increasingly adopting recycling as a method of managing municipal waste. Whether publicly or privately operated, the process diverts a significant percentage of municipal solid waste from being disposed. The list of potentially recyclable materials is long, and it continues to grow as technological developments enable more materials to be recycled.

The Three Rs Principle

The 3 'R's are the most viable alternatives in the manner in which we can help in managing waste.

Reduce what you do need

Reuse what you cannot reduce

Recycle what you cannot reuse

Reduce - Reducing the waste stream is the most significant of all the options to manage waste. The problem of waste management is reduced to the extent we are able to reduce the generation of wastes. Thus, we can conserve our resources and reduce disposal costs and pollution. We can start by analyzing what we throw away at home. This will include thinking about the goods and services we buy and the activities we support, and in what ways do they contribute to the solid waste problem.

Reuse - The process of reusing is started with the assumption that the used materials that flow through our lives can be a resource rather than refuse. We can 'reuse' many materials in their original form, instead of throwing them away, or pass them on to others who can use them. Once we have our minds set on the fact that we can use trash positively, we can begin to brainstorm and generate ideas. For this, we need to analyze our activities very closely and pick out things that can be reused. Reusing products or materials also helps in generating less waste, which again reduces the cost of managing solid waste and conserves natural resources.

Recycle - Recycle is a stage where the reusable material is remade into either the same or new product. So, it is a resource recovery method involving separating, collecting, processing, marketing and using material that would have otherwise been thrown away. However, recycling must be based in the overall production and consumption cycle. It should not be looked upon as an activity that can exist on its own merely as an alternative disposal means. In fact, it is one component of integrated waste management. The entire recycling process is a dynamic mechanism in itself. With increase in supply of materials, manufacturing facilities, in both organized and unorganized sectors, have emerged to find uses for recyclable materials. This subsequently has created jobs or livelihood for a number of persons. This also helps in reducing the quantity of wastes going to landfills.

Toxics Link's Decentralise Solid Waste Management Model

There are several methods through which the objectives of decentralised waste management could be accomplished. Non

Governmental Organisations across the world have shown how these methods can be operationalised. For example, the initiatives undertaken by Toxic Links (Delhi based environmental NGO) have primarily adopted the community-based model with the view of developing the environmentally and economically sustainable decentralised solid waste management. Toxics Link, following the zero waste management strategy facilitated many residential colonies and education institutes in Delhi and out side Delhi. Defence colony is one of the zero waste management colonies. The details of the model intervention have been discussed in the next few paragraphs.

Project Background

"Success comes with hard work". Toxics Link's initiative with strong support and participation of the residents of A-block, Defence Colony speaks volumes about the effort and perseverance that has gone into the success of the community-based solid waste management. Defence Colony falls under the Central Zone of Municipal Corporation of Delhi. Unsatisfied with the existing system of solid waste management, the Resident Welfare Associations(RWAs) decided to take the responsibility to make its community a cleaner and hygienic place to live in.

The intervention began in the year 2005 in the month of June when Toxics Link received a note of interest from Residents Welfare Association (RWA), A-Block, Defence Colony for establishing an environmentally sound solid waste management in their community. There are altogether more than one thousand households (1000 HHs) in the A-Block of Defence Colony that belong to the category of middle-high-income group. Collection of waste from every household was in place but the total collected waste was disposed off into the municipal *dhalaos* located within the community by the waste collectors. The eight waste collectors earn their income by collecting a monthly user fee of Rs.35 from every household. Each waste collector caters to an average 200-225 HH and the monthly salary is supported by his or her own collection. When the project began the sanitation condition in the locality was poor with overflowing municipal bins / dhalao and stinking garbage dumps largely due to irregular cleaning and neglect by the municipal authorities and workers. There was no system of segregating the dry and wet waste.

Implementation Approach

In decentralized solid waste management, community is the most important stakeholder in bringing sustainability to zero waste zone initiative. Therefore, it was made clear to the RWA, A-Block, Defence Colony that unless they geared up to drive the project, such initiative would only achieve partial success. The meetings were organized on the underlying dynamics of the initiative and discussed the respective roles and responsibilities of the various stakeholders.

Four compost pits were constructed in JCO club with the size of 7x5x4 cubic feet. Initiation of such a system requires space for locating compost pits and procurement of the required infrastructure. The RWA of this block has made an appreciable effort by constructing the compost pits and coordinating with the waste collectors. Recognizing the importance of the occupational safety of the workers, Toxics Link addresses the issue and committed to ensure that the process does not cause health hazard to the workers.

Inauguration of Community Based Composting at Defence Colony A - Block

The strength of the initiative lies on the active involvement of the RWA members who have kept a strict vigilance on the waste collectors and the compost turners. The residents were educated through the meetings and distribution of information materials to get their active involvement in achieving the success. Progressive and consistent involvement of the core organization and other stakeholders lead to the sustainability of the project to create zero waste community.

Capacity building of waste collectors

Turning of garbage by one of the worker

In this context, the residential colony is the core institution. Once Toxics Link was ensured of the involvement of the RWA as well as the residents of the colony.

Project was initiated with about 1000 households in Block A and later on it was expanded to cover about 4000 households from all Block of Defence Colony. This involves door-to-door collection of waste, primary and secondary segregation and community based composting. The project adopted the following approaches:

1. To provide daily door-to-door collection service using cycle carts operated by teams of waste collectors. The waste pickers are provided with uniforms and safety equipment. Household are trained to segregate waste at source into organic and inorganic waste;
2. The waste is also segregated at composting site into compostable, recyclable waste, and inert waste. Strict monitoring for segregation is maintained in composting, after which the waste is enriched to make organic fertilizer. Recyclable material is sold to junk dealers (kabadiwala) and inert wastes are taken to the municipal dumpsite;
3. The community based solid waste management is a profit making enterprise in terms of three streams of incomes: door to door collection fees from households, sale of enriched compost, and sale of recyclables to junk dealers. This supports in sustaining the project.
4. The approach provides several benefits to all participating stakeholders. For local governments, it minimizes waste transport and disposal costs. For the participating households it provides improved collection service. For the waste collectors it provides higher and regular income and better working conditions;
5. Invoke active participation and ownership of the community-based solid waste management activity of RWAs. Launching of awareness campaign for residents with regard to primary segregation of waste through

regular issue of circulars by the RWA supplemented by door-to-door visits by the Community Activists from the Toxics Link;

6. Orientation cum training programmes for the waste collectors. Four such orientation-cum-training workshops were conducted where the waste collectors learned about waste segregation and methods of composting. Special attention was given to occupational safety and methods of social interaction with their client households.

Benefits of Decentralized waste Management

Consequently, by the collaborative efforts of community and Toxics Link, around 30 per cent households have started handing over segregated waste to the waste collectors while the remaining mixed waste is segregated by the waste collectors to recover the recyclable and reusable materials for sale. Kitchen waste is then taken to the compost pits. This has to a great extent helped in reducing the dependence on MCD *dhalaos*, leading to diversion of waste from landfill. Monthly collection fee from the residents supports the salary of the waste collectors. It is encouraging to learn that the waste collectors earn an additional income by selling the valuable dry wastes to kabariwalas. The whole exercise has led to least amount of waste being actually dumped into the MCD dhalao. The initiative shows a significant impact on the cleanliness and the aesthetic sensibility of the community. Some of the major impacts have been discussed below.

Recovery of Resources Through Composting

In the beginning due to the availability of land, the composting was initiated with the wet waste collected from 500 households of the 1000 households in A-Block of Defence Colony. Four workers were initially involved in the composting process. According to the survey conducted by Toxics Link, it was found that average daily generation of waste from each household was approximately 0.85-1.05 Kg comprising of both wet and dry waste.

Table. 1: Details of Generation of Wet Waste

No. *of households*	*Wet Waste / HH / day(grams)*	*Total Wet waste in Kg / day*	*Total Wet waste in tonne/day*	*Total Wet waste in tonne/ annum*	*Waste diverted from landfill/ dayKg*
500	600-700	300-350	0.3-0.35	110 - 128	300-350
1000	-Do-	600-700	0.6-0.7	219-256	600-700
4000	-Do-	2400-2800	2.4-2.8	876-1022	2400-2800

Source: Field Survey

Therefore, total waste generated was making its way to the landfill was approximately one tonne/day. The average recyclables produced by every household was 250-350 g per day. From the total one tonne of waste generated per day approximately 750 kg of wet waste is composted every day. The details of generation of wet waste have been presented in table 1.

Revenue and Employment Generation

Three sources of revenue generation through the project on community based solid waste management were observed;

a. Sale of recyclables collected from households through primary and secondary segregation

b. Door to door collection fee received by the waste collectors directly from the client households;

c. Indirect saving on transportation of waste as 70-80 percent of household waste is being managed through the composting and the recycling.

It was noted that about 250-350 Kg of recyclables was being collected from 1000 households every day. The sale of recyclables generated Rs. 50-75 per waste collector. In total, there is a revenue generation of Rs.400-600 per day only by selling the recyclables collected from 1000 households. In means, on an average each waste collector sells recyclables worth of Rupees 1500-2250/- per month. Each household pays user fee of Rs. 35/- pm for the door to door collection of waste. The revenue generated per month through collection fee from the 1000 households is Rupees 35, 000. The fee is collected directly by the waste collectors. Thus, besides surety of employment and regularity of earning to the waste collectors each one of them earned an average of rupees 4,375/- per month.

Table- 2: Details of Recovery of Recyclables

No. of households	*Recyclables /HH /day (g)*	*Total recyclables*	*Total recyclable (tonne)/day*	*Total recyclable (toone)/ annum*
500	250 -350	125-175	0.125-0.175	45.625-63.875
1000	-Do-	250-350	0.25-0.35	91.25-127.75
4000	-Do-	1000 - 1400	1.0 - 1.4	365.0-511.0

Source: Field Survey

If the earning from the sale of recyclables is also to be included then each waste collector earns about rupees 6,000 per month. The details of the recyclables have been given in table 2.

Landfill diversion

It was found that approximately 600-700 Kg of waste per day is diverted from landfill through this small effort. Management of wet waste by community in A-Block (1000 households) helps divert 600-700 Kg of waste from the landfill. With the upscaling of this programme in other blocks of Defence Colony (4000 households), this amount has risen to about 2400-2800 Kg (2.4-2.8 Tonnes) of wet waste per day. If recyclable materials were also included the total diversion works out to in the region of 2650-3150 Kg (2.6-3.0 Tonnes) of municipal solid waste. This way it has reduced the storage and transportation cost to the municipal corporation. The details of the landfill diversion are presented in the table 3.

Table 3: Details of Manageable Waste

No. of household	*Total wet waste day (kg)*	*Total recyclable day (Kg)*	*Total Waste day (Kg)*	*Waste diverted from landfill/ day*	*Approx. waste diverted from landfill/ annum*
1000	600 -700	250 – 350	850 - 1050	850 – 1050	346.75 tonnes
4000	2400–2800	1000-1400	3400-4200	3400–4200	1387.0 tonnes

Source: Field Survey

Cost Analysis for Decentralised Waste Management Model

In the beginning the wet waste from the 500 households out of 1000 households was taken into the secondary segregation and composting. For initiating the composting process, infrastructure needs to be established first. The model takes into account the initial establishment cost, which includes procuring of rickshaw and related equipments, cost of construction of compost pits and its maintenance and the cost involved in training and building the capacity of the waste collectors.

While the establishment cost is one-time, there are other recurring costs, which need to be supported throughout. These include the cost involved in purchasing Effective Micro-organism (EM) solution and labor cost for monitoring the composting process. The monthly maintenance of the rickshaw and the salary for the workers are also recurring costs, which need to be borne by the owner of the initiative.

The revenue generated through the programme as given here can meet the monthly recurring cost and can reach a break-even (revenue generated covering the establishment cost) point in certain time period. There are specifically three modes of revenue generation, collection fee from the residents or the user fee, the sale of compost and the recyclables. In A-block, the waste collectors get the user fee from the residents in the form of monthly salary. The waste collectors also own the recyclables for sale. The detail of the cost involved in creating a solid waste management model at community level has been presented in table 4.

Table 4: Cost Analysis for 500 HHs in Middle-Income Group Community

Establishment Cost (In Rupees)		*Recurring cost Monthly (Rupees)*		*Revenue generation Monthly (Rupees)*	
Elements	*Cost*	*Elements*	*Cost*	*Source*	*Amount*
Two rickshaws @ Rs.6, 000/ rickshaw	12,000	4 Workers @ Rs 4375/ worker (Distribution of the monthly user fee)	17, 500	User fee @ Rs.35/ HH	17,500
Digging of 4 pits required for sustaining the programme @ Rs.750 pit of (6x4x3) cubic ft in size	3,000	2 bottle of EM solution @ 210/bottle & 4 Kg Jaggery @ Rs.24/Kg	420+96 = 516	Sale of compost	3,000
Construction materials for 4 pits	12,000	Monitoring cost	500	Average sale of recyclables @Rs.50 worker/ day (Additional earning of waste collectors	6,000 (Rupees 1,500 per worker)
Pit construction labor cost 4 @ Rs.500/ pit	2,000	Labor cost for the worker (Turning & Sprinkling)	500		
Equipment cost (barrel, spade, fork, ganti, bucket, mug, spring balance, jute and plastic sheet) Apronglovesed pair boots	5,000	Sieving cost @ Rs.500/- month	500		
Capacity building of workers IEC materials	5,000 5,000	Incidental cost	500		
Total	44,000		20,016		26,500

Source: Field Survey

- *This model is for the middle income group community where the private door to door waste collection is already into practice and the monthly user fee and income through recyclables directly goes to the waste collectors.*

Monthly income of the workers:

- Average monthly income to four workers: Rupees 17, 500 + 6, 000=Rs 23, 500
- Per worker average monthly income: Rupees 23, 500/ 4= Rs 5, 875[1]

Break-even Analysis:

In order to derive a sustainable programme, it is important to project the break-even point.

The break-even point is the point where revenue generated is exactly equal to costs. At this point, no profit is made and no losses are incurred.

Break-even analysis is a useful tool to study the relationship between fixed costs, variable costs and returns. A break-even point defines when an investment will generate a positive return. Break-even price analysis computes the price necessary at a given level of production to cover all costs. To explain how break-even analysis works, it is necessary to define the cost items.

Fixed costs, incurred after the decision to enter into a business activity is made, are not directly related to the level of production. Fixed costs include, but are not limited to, depreciation on equipment, interest costs, taxes and general overhead expenses. Total fixed costs are the sum of the fixed costs.

Variable costs change in direct relation to volume of output. They may include cost of goods sold or production expenses such as labor and power costs, feed, fuel, veterinary, irrigation and other expenses directly related to the production of a commodity or investment in a capital asset.

Case One:

- Since the RWA is the owner of the programme and has invested towards establishing the programme
- Total establishing cost: Rupees 44, 000
- Monthly recurring cost to RWA: Rupees 1, 016 (EM, jaggery and incidental cost)

- Monthly revenue generation by selling the compost: 3, 000 (Only compost goes to RWA)

So, The break-even period will be:

B-E = F / (S - V)

where:

B-E = break-even point (number of months),

F = total fixed costs,

V = variable costs per unit of production,

S = savings or additional returns per unit of production

Based on the above calculation, break-even point can be estimated.

B-E =44, 000 / (3, 000– 1, 016) = 22.18 Months

Therefore, it will take 23 months to cover the total costs invested initially for establishing the system.

Case- Two

If Incidental cost is borne by the workers:

- Per worker incidental cost: Rupees 500/4=125
- Then, per worker monthly income: Rupees 5, 875-125=5, 750
- Monthly recurring cost to RWA: Rupees 516 (EM & jaggery cost)
- Monthly revenue generation by selling the compost: Rupees 3, 000 (Only compost goes to RWA)

So, the breakeven period will be:

B-E = F / (S - V)

B-E = 44, 000/(3000-516)

B-E = 44, 000/2484 = 17.71 months

Therefore, it will take 18 months to cover the total costs invested initially for establishing the system.

Areas of Concern

1. RWAs in the locality along with their federation have remained very active throughout during the project. But participation of the residents in the solid waste management activity remains low and poor;

2. In spite of residents' higher standards of living, high income and educational levels the idea of segregation at source has remained poor. This was largely due to the fact that a large proportion of the households had full time or part time servants and maids looking after the sweeping and kitchen activities. The need is to train this segment along with regular awareness campaigns within the locality;

3. Availability of land for secondary segregation and composting remains one of the most crucial factors underlining the project on community-based solid waste management. The role of Municipality, in this regard has been rather limited and discouraging;

4. Marketing of compost as a source of earning is equally crucial for the sustainability of the community based solid waste management and composting activities. The RWA has not been in a position to market its compost. The off-take from the residents is low because the non-competitive pricing;

Recommendations

1. Since the solid waste management is a people's programme, the implementing agencies should adopt the participatory approach during the entire course of implementation, which can create a sense of ownership among the community, which is the critical need for the sustainability of the programme. As the RWA is the reprehensive of the community and also taking decisions on behalf of the residents. The need is to further diversify and decentralise various activities of the association based on tasks, themes and areas of community interest;

2. Local schools should be involved. The schools can work as a bridge in dissemination of information from the functionaries to the families and the entire community;

3. Institutionalization linkage between municipality and the community is critical need for sustainability of the programme;

4. There should be a provision of the incentives for the communities who are coming forward for the waste management programme in terms up scaling the SWM;

5. For the self-sustainability of the programme, ownership must lie upon waste collectors, community (RWA, CBOs, or Committee) helps in creating sustainable interest among them;

6. The horticulture department and other departments of the Union and State governments should give priority to RWA in purchase of organic fertilisers, manure etc. This would help in the sustainability of the project;

7. In Defence Colony A-Block, it is found that a considerable amount of recyclables is generated, which has good sale value. The waste collectors take charge of the recyclables and earn additional income by selling these items. If these items can be collected in a systematic way by the RWA, it can supplement the monthly recurring cost. Not only the saleable recyclables but also sale of compost is also another source of revenue generation, which contributes to meet out the recurring cost. It can also be seen that once the association takes charge of the program, cost of external monitoring visit is not required. It is very essential to take into account the recovery from sale of recyclables and compost since the user fee alone cannot support the monthly recurring cost. Therefore, it is recommended that when any association initiates such a programme, it should be planned in a way that recyclables are handed over to them.

End Note

- In order to make the project self-sustained the cost like (Monitoring, Labor and Sieving) should be born by the waste collectors as all earning through monthly user fee and by selling the recyclables directly goes to the workers or they themselves should perform these activities
- On the other hand, since RWA is the owner of the manure so they will be bearing the monthly cost of EM solution & jaggery and Incidental cost.
- Estimated cost of organic cost: Rupees 10/Kg.

References

1. Toxics Link, (2005), "Clean Community - A Comprehensive guide to effective solid waste management, Toxics link, New Delhi

2. Ministry of Urban Development (2000), "Manual on Solid Waste Management (Ist ed)", Prepared by the expert committee constituted by the Ministry of Urban Development, Government of India, January 2000.

3. Daniel Hoornweg, Laura, Thomas and Lambert, Otten, "Composting and its applicability in developing countries" 8th working paper Series,. Published for the urban development division, The World Bank, Washington DC.

4. http://www.ext.colostate.edu/PUBS/farmmgt/03759.html

5. http://connection.cwru.edu/mbac424/breakeven/breakeven.cfm

6. Gutierrez,1 P.H. and N.L. Dalsted *(September 1992, Reviewed October 2004)*, "Break-Even Method of Investment Analysis" Agricultural and Resource Economics. 9/92. Reviewed 10/04)

Chapter 15

Domestic Solid Waste Management: A Case Study of Gaziabad District of U.P.

Dimpal Vij and Babita Aggarwal

INTRODUCTION

Waste is everybody's business. We all produce unwanted byproducts and residues in nearly everything we do. Though generation of solid waste is not a new phenomenon, it has acquired a dangerous status of being third pollution after air pollution and water pollution with progress in industrialization, urbanization and population explosion. Earlier waste management was hardly considered to be an issue of concern because the major constituents of solid waste were domestic wastes and agricultural residues which are both biodegradable and it could be disposed off in an environmentally safe manner within the generation premises but with passage of time due to advancement, increasing consumerism and changing lifestyles of people, management of solid waste has come as a problem so much so that not only government but also residents at their own level are trying to solve this problem. But before coming to the management of domestic solid waste let us come to the basics first.

Solid Waste[1] (SW) can be defined as the material that no longer has any value to the person who is responsible for it and is not intended to be discharged through a pipe. It does not normally include human excreta. It is generated by domestic, commercial, industrial, healthcare, agricultural and mineral extraction activities and accumulates in streets and public places.[1] The words 'garbage', 'trash', 'refuses', and 'rubbish' are used to refer to some forms of Solid Waste. Solid Waste produced by human activities can be divided into four groups. (see table 1)

1. **Agricultural Wastes** - In India, the main sources of agricultural wastes are wheat , paddy and maize straw, sugarcane trash, rice and wheat bran, maize cobs, leftovers from pulses etc. There has been a great increase in the generation of crop residues and allied wastes.

2. **Municipal and Domestic Wastes** -- There are different sources of Municipal Solid Waste (MSW): domestic, market community facilities etc. The amount of MSW generated is dependent on the public habits which can vary from city to city and even among towns. In India, per capita MSW production in metropolitan cities is significantly high in comparison to the towns and villages. Average per capita generation of waste is about 0.4kg/capita/day in India.

Domestic Solid Waste (DSW) is waste generated from households or residential units. They generally consist of wet or organic waste and inorganic waste portions. Wet wastes are usually made up of food and kitchen waste and yard trimmings or other garden waste. Inorganic waste usually consist of paper, corrugated cardboard, plastic, glass, wood, ferrous and non-ferrous metals that are found in many household products. Most often these inorganic wastes can be recycled.

3. Industrial Wastes - Huge amount of industrial Solid Waste are usually produced by different industries. The estimated SW of industrial origin contribute only 10% of the total wastes generated, the bulk is liquid.

Table 1: Solid Wastes produced by Human Activities

S.No.	Typesof Human activities	Example of wastes liberated
1.	Agricultural	Plant remains, processing wastes, animal wastes.
2.	Domestic	Garbage, rubbish, wastes produced at home from cooking etc.
3.	Municipal	Street sweepings, wastes from schools, offices and otherInstitutions.
4.	Industrial	Wastes produced by mining operations, manufacturing andConstruction works.

(Source: Misra and Mani 1993)

In this study we are dealing only with Domestic Solid Waste (DSW) which is posing a great threat not only to health of the residents, their pets and stray animals but also killing the aesthetic beauty and creating environmental hazards in th posh colonies and apartments of urban and suburban areas. With the changing lifestyles this problem is also percolating down in rural areas and certainly going to be increased in the coming years.

Current Status of Domestic Solid Waste (DSW)

Earlier Domestic Solid Waste was not a problem because maximum DSW was of biodegradable nature i.e kitchen waste like vegetables and fruit peels. People were used to feed this waste to their own or neighboring cattle's or the stray animals because this was mandatory in our culture that no part of food should be wasted. It should be either consumed by ourselves or given to the pets or stray animals. This also showed co-existence part of our cultural values that teaches us the dependence of both animals and human beings on each other. And this was the way to do away with maximum (60-70%) domestic solid waste which is in the wet form. Even today in rural areas this domestic solid waste is not such a grave problem due to our inherited cultural values which our forefathers were followings from last many decades even being illiterates. But over the last few years with increasing consumerism, changes in our food habits and lifestyles (eating and drinking everything in disposable items) and changing marketing policies like to pack the products in attractive packaging like cans, aluminum foils, plastics, poly-bags etc. has aggravated the problem manifold. Today situation is this that even for the small occasions at home likes Puja, family or peer group gatherings, festivals or birthday celebrations we prefer to use disposable items rather than reusable utensils of home using disposable items considered as a status symbol. And this culture is increasing day by day not only in urban areas but also in suburban and rural areas resulting in great heaps of garbage in colonies, on roadsides and other emptied areas. Being non-biodegradable these poly-bags and plastic glasses remains at these places for many days, feeding to the stray animals that not only feed on it but scatter these here and there in the form of litter. This in turn creates many health hazards to them because these poly-bags accumulate in their body and ultimately results to their death. When human being consume the meet of such animals they also get many diseases unknowingly like swine flu, bird flu etc.

Our Attitude Towards Waste

At present in India the attitude of the people regarding this waste disposal is such that people feel proud in disposing this waste in front of others houses (NIMBY's Attitude) or on the roadside or even on the wide divider of the road as we found at Badshahpur, Haryana as if to exhibit that they are doing something of greatness to show the passers-by. The magnitude of the problem can be felt by the fact that when one enters in a colony, town, city or even in a state one is greeted with this waste dumps lying on the roadside giving us a feeling of one nation no matter to what caste, creed or religion do we have, but we have the unique same habit of domestic waste disposal of littering it on the roadside or in any emptied area wherever we find all over the country be it Haryana, Punjab, Gujrat, UP or some other state.

If we see this problem statistically we find that nowadays per capita waste generation is increasing by 1.3% per annum while in urban population this increase is between 3-3.5% per annum. Though the yearly increase in waste generation is around 5% per annum.

Presently India produces 42.0 million tons of municipal solid waste annually while per capita waste generation varies from 200gm to 600gm per capita per day. Average generation rate is at 0.4 kg per capita per day in 0.1 million plus towns. If we see the collection capacities of the civic bodies we find that collection efficiency ranges between 50% to 90% of the total solid waste generated. Of the total amount spent by the civic bodies to clean this waste 60-70% amount is spent on the collection, 20-30% on the transportation and hardly any amount left for treatment or disposal of waste and this crude dumping of waste in most of the cities is creating various health and environmental problems.

Problems Due to Domestic Solid Waste

Problems occurring due to domestic solid waste can be divided into two parts : Health problems and Environmental and other problems.

Health problems- As the waste is spread on streets, road sides and other public places it attracts flies, insects and other parasites causing many vector born diseases like malaria, dengue kala-Azar, fever, loose motions etc.

Environmental and other problems - As the waste is normally wet and if remains unattended for many days, it decomposes and gives a bad smell and it becomes very difficult for the passers-by and nearby residents to remain in that environment. The fresh air gets polluted and also the aesthetic pleasure of the colony and roads which not only gives a bad look but also shows our behaviour towards our own waste. Sometimes when this waste enters into the water, it pollutes water bodies and give rises to many water born diseases. When this waste goes to nearby drainage system it clogs the drainage system resulting in overflowing of water of these on the roads and colony and this creating various problems.

Action Plans to Solve This Problem

If we break up the total domestic solid waste of an average Indian family we find that maximum 60-70% of the waste produced everyday in a family is of biodegradable nature i.e. vegetables and fruits peels or food items residues. 20-30% is of recyclable nature i.e. plastic bags, cardboard packing bags, glass bottles, aluminum or iron items etc. only less than 5% is of hazardous toxic nature i.e. electronic items, children toys, batteries etc.

For 60-70% kitchen waste of biodegradable nature it is possible to devise solutions both at individual and community level to deal with this problem of MSW. Two case studies from Gaziabad District of UP have demonstrated how this is possible.

Individual Level - Dr. Abhay Pal, a resident of posh vasundhara colony of Ghaziabad city was very perturbed to see the overflowing of garbage bins of his colony. Even this garbage (being scattered by the stray dogs in the colony and roadsides) not only spreading the foul smell in the colony but also marring the aesthetic beauty of the colony and decided to do something at his own by asking his neighbour families not to throw the kitchen waste into dustbins but donate to him in the morning when he was going to them to collect it. After collecting the waste he was putting them in two big earthen pots into 1:5 ratio of garbage and water and after around 45 days organic manure was ready for the plants and this manure was being given to his two hundred plants of his house and to the colony free of cost.These plants were got healthier day by day in comparison to those neighbourhood families plants who have been buying inorganic chemical fertilizers from the market spending a handsome amount of money. In the beginning people thought that he is a mad person but when gradually people saw the benefits of his initiative they started the same process of making organic manure out of kitchen waste free of cost thus saving lot of money spent on buying chemical or organic fertilizers.. When this news was published in Hindustan Times Ghaziabad Live section and people who read this news also get motivated and started the same process of making manure in many other colonies.

Community Level -Second case study is about Maliwara colony, a very old colony in the heart of Ghaziabad city with 2000 houses in 10 lanes having narrow roads. This project was started by members of Yamuna Action Plan of UP government as a pilot project in 3 lanes of Maliwara with 150 houses. Initially 300 buckets were distributed in these 150 houses one yellow bucket for inorganic waste and another green bucket with for collection of kitchen waste. Residents especially women were provided with awareness and prepared to segregate inorganic waste with kitchen waste. Public response was overwhelming. Tired of seeing garbage lying here and there in the colony l people were very enthusiastic to get rid of this waste everyday at 9.00a.m when a private sweeper comes to their door to collect the kitchen waste. After collecting from the houses he brings this waste to the 3 compost "pits" with 3ft deep and 8 ft wide digged at the back of the colony. The pits are covered and small cervices are used as inlets for air and within 45days compost is ready which is sold to the nurseries at a cost of Rs. 2 per kg. This colony produces 1 qt. of kitchen waste per day but now that does not go to garbage dumps or spreads in the colony here and there. Out of this 1 qt. 10 kg manure is readied and sold.

The response is so overwhelming that the residents of other 7 lanes are also eager to take such action plans in their lanes and make their lanes garbage free and thus adding to aesthetic beauty and health status of the residents.

This is not exhaustive but only an indicative list. Many such small plans are being run in many of the colonies of many of the states and may vary according to the needs of the residents. The same programme can be run at the village level with the help of the village panchayat. For this volunteer villages should be choossen and if family participation increases more than 2000 families. A biomass plant can be established by the government in the colony or the village to generate the power but the condition is this the power generated through this plant should be used by the colony people or the villagers who contribute their kitchen waste to make the power and the slogan should be 'My Waste My power' 'My Waste My Manure'.

For the remaining 20-30% waste of non-biodegradable nature (clubbing all domestic waste excluded from kitchen waste) can be segregated at the home level and put in different buckets and bags and can be disposed off by giving it time to time to the Kabadies. For those families who don't have time or storage place to segregate it, store it and sell it, it can be directly given to their daily sweeper who at his own level can segregate it and sell it to different shopkeepers of Kabad and earn the money at the end of the day to support his family. From these shopkeepers it can reach to respective recycles factories or centers and thus turns up in making new items.

In these action plans, formal bodies should not be be involved and if they are involved the efficiency of the programme will suffer. Government support is required when this action plan is being run on a large scale (i.e. >2000 families) and we have enough waste to generate biomass energy. So at that level government can help to establish the biomass plants to generate power from this waste. But this power should be given to the colony or village itself from whose garbage this power is being generated otherwise the plan will not be successful.

Benefits of these Action Plans are:

- Garbage free colony;
- Lesser health hazards especially communicable and vector born diseases
- Savings in out of pocket helath expenses;
- Getting organic manure for their plants and agriculture either free of cost or at a very nominal rate;
- Providing employment to all in the chain of recycles procedure;

- Lessening Environmental problems like air pollution, soil pollution, water pollution etc.;
- Adding to the aesthetic beauty of the colony/village.

Conclusion

Thus with these various action plans which we have given above we can solve the problem of domestic solid garbage which is littering here and there in the whole country and giving us the image of **'Dirty Nation'** and can turn this waste into gold providing benefits to each person who one way or other is involved in turning this waste into the productive use. In this way we can clean our colonies or villages and can make them **'Garbage free zone'**.

Bibliography

Ackerman, Frank (1997), 'Why Do We Recycle?' Washington D.C, Island Press.

Akolkar, A.B. (2005). Status of Solid Waste Management in India, Implementation Status of Municipal Solid Wastes, Management and Handling Rules 2000, Central Pollution Control Board, New Delhi.

Asnani, P.U. (2004). "United States Asia Environmental Partnership Report", United States Agency for International Development, Centre for Environmental Planning and Technology, Ahmedabad. (2005). Technical Committee Report, West Bengal SWM Mission 2005, Government of West Bengal, Kolkata.

Atri, Said and Thomas Schellberg (1995), 'Efficient Management of Household Solid Waste: A General Equilibrium Model', Public Finance Quarterly, 23(1), January, 3-39.

Banerjee Sharmila (2001), 'Valuation of Environmental Benefits and Costs', in Rabindra N.Bhattacharya (ed) 'Environmental Economics', OUP, New Delhi, 125-161.

Bartone, Carl R., J. D. Bernstein and F. Wright (1990), 'Investments in Solid Waste Management: Opportunities for Environmental Improvement', Policy Research and External Affairs Working Paper No. 405, The World Bank, Washington D.C.

Baumol, William J. and Wallace E. Oats, (1988), 'The Theory of Environmental Policy', 2nd Ed., New York, Cambridge University Press.

Beukering, P. V. (1997), 'Waste recovery in Bombay: a socio-economic and environmental assessment of different waste management options', Third World Planning Review 19(2): 163–187.

Beukering, P.V., S. Madhushree, R. Gerlagh, and V. Kumar (1999), 'Analysing urban solid waste in developing countries: a perspective on Bangalore, India', CREED working paper no. 24, International Institute for Environment and Development, London UK.

Calcott, Paul and Margaret Walls (2000), 'Can Downstream Waste Disposal Policies Encourage Upstream Design for Environment?' American Economic Review, 90, No.2.

Coad, Adrian (ed.) (1997), Lessons from India in Solid Waste Management, Water, Engineering and Development Center (WEDC). Loughbrough university, Leicestershire UK.

Colon, M. and Fawcett, B., "Community-based Household Waste Management: Lessons Learnt from EXNORA's 'Zero Waste Management' Scheme in Two South Indian Cities", Habitat International, Elsevier Publication, vol 30;issue 4:dec 2006,pg. 916-931

CPCB (Central Pollution Control Board) (2000), Management of MSW, Ministry of Environment and Forests, Government of India.

GOI (2003). Report of the Technology Advisory Group on Solid Waste Management, Government of India Publications, New Delhi.

India Today (1994), 'Our Filthy Cities: Can We clean Up the Mass?', October 31, 62-79.

Kumar, S. and Gaikwad, S.A., "Municipal Solid Waste Management in Indian Urban Centres: An Approach for Betterment", Urban Development Debates in the New Millennium, Edited by K.R. Gupta, Atlantic Publishers & Distributors, New Delhi, pp.100-111, 2004.

Maimone, M. (1985), 'An application of multi-criteria evaluation in assessing MSW treatment and disposal systems', Waste Management and Research 3: 217–231.

Misra S. G. & Mani D. (1993) In: Pollution through Solid Waste. Ashish Publishing House, New Delhi

MOUD Report (2005). Management of Solid Waste in Indian Cities, Ministry of Urban Development, Government of India, New Delhi.

(1999). What a Waste: Solid Waste Management in Asia, World Bank, Washington, DC.

MOUDPA (2000). Manual on Solid Waste Management, Ministry of Urban Development and Poverty Alleviation, Government of India Publications, New Delhi

NIUA (National Institute of Urban Affairs) (1999), Solid waste management: , improvement initiatives in selected cities of India, Research study series no.75 New Delhi.

R.P.V. Beukering, M. Verma, P.P. Yadav, and P. Pandey (1999), 'Integrated modeling of solid waste in India', CREED working paper no. 26, International Institute of Environment and Development, London.

SC (1999). Report of the Supreme Court Appointed Committee on Solid Waste Management in Class I Cities in India, Supreme Court of India, New Delhi.

UN (United Nations) (2000), 'State of the environment in Asia and the Pacific 2000', United Nations, New York.

World Bank (1997a). What a Waste, World Bank, Washington, DC. (1997b). Per Capita Solid Waste Generation in Developed Nations, World Bank, Washington, DC.

www.hindustantimes.com

www.ijoem.com

Notes

1. Indian Journal of Occupational and Environmental Medicine - August 2008 - volume 12 issues 2.
2. Misra S.G. & Mani D.(1993) In pollution through solid waste.Ashish Publishing house, New Delhi

Chapter 16

Municipal Solid Waste Management: A Case Study of "The City of Taj"

Roopam Srivastava and Anurag Srivastava

Theoretical Framework

Rapid and extensive industrial development, unplanned urbanisation, regular flow of persons from rural to urban areas and improper and inadequate action of the authorities entrusted with the work of pollution control and environmental protection have largely contributed to unhealthy and degraded environment. Unplanned and alarming rate of urbanisation has given rise to many environment related problems, such as problem of health and hygiene, sewage, disposal of solid waste, air, water and land pollution, slums, housing, basic amenities and others. Apart from air and water pollution problems, most gigantic problem which modern cities and industries are facing is the solid waste disposal problem. It is predicted that by the year 2015, 53 percent of the world population will be city dwellers and it is the third world that will face urban doom. The management of solid waste problem is the by-product of this urban doom. Management of solid waste includes production, transportation, disposal techniques and its treatment.

Solid waste management

The US Congress, in the 1976 Resources Conservation and Recovery Act defined 'solid waste' as— 'Any garbage, refuse, sludge from a waste treatment plant, or air pollution control facility and other discarded material, including solid, liquid, semisolids or contained gaseous material resulting from industrial, commercial, mining and agricultural operations and community activities.'

Management of solid waste is associated with the control of generation, storage, collection, transfer and transport, processing, and disposal of solid wastes in a manner that is in accordance with the best principles of public health, economics, engineering, conservation, aesthetics, and other environmental considerations. In its scope, it includes all administrative, financial, legal, planning and engineering functions involved in the whole spectrum of solutions to problems of solid wastes thrust upon the community by its inhabitants.

Municipal Solid Waste

The Municipal Solid Wastes (Management and Handling) Rules, 2000 has defined 'municipal solid waste' as 'commercial and residential wastes generated in municipal or notified areas in either solid or semi-solid form *excluding* industrial hazardous wastes but includes treated bio-medical wastes.'

Types of Municipal Solid Waste

Table 1: The Sources of Municipal Solid Waste

Sources	Examples
Residential	Single family homes, duplexes, town houses, apartments
Commercial	Office buildings, shopping malls, warehouses, hotels, airports, restaurants
Institutional	Schools, medical facilities, prisons
Industrial	Packaging of components, office wastes, lunchroom and restroom wastes (but not industrial process wastes)

Municipal Solid Waste Management

Municipal Solid waste management involves the application of principle of Integrated Solid Waste Management (ISWM) to municipal waste. ISWM is the application of suitable techniques, technologies and management programs covering all types of solid wastes from all sources to achieve the twin objectives of (a) waste reduction and (b) effective management of waste still produced after waste reduction.

In the Municipal Solid Waste Management the major issues to be considered are:

- Increasing waste quantities
- Wastes not reported in the national MSW totals
- Lack of clear definition for solid waste management terms and functions
- Lack of quality data
- Need for clear roles in state and local government
- Need for even and predictable enforcement regulations and standards

Functional Elements of Municipal Solid Waste Management

The activities associated with the management of municipal solid wastes from the point of generation to final disposal can be grouped into the six functional elements:

(a) Waste generation;

(b) Waste handling and sorting, storage, and processing at the source;

(c) Collection;

(d) Sorting, processing and transformation;

(e) Transfer and transport; and

(f) Disposal.

Legal and Regulatory Aspects

The municipal solid waste shall be managed and handled in accordance with the compliance criteria and procedure laid down in Schedule II of The Municipal Solid Wastes (Management and Handling) Rules, 2000. Schedule II has provided the compliance criteria for the management and handling as follows

1. Collection on municipal solid waste

The littering of MSW shall be prohibited in cities, towns etc. and organize house-to-house collection on regular and pre-informed timing by using bell, singing, devise collection of waste from slums, squatter areas, slaughter house, fish markets, fruits and vegetable, industrial waste, horticultural and construction or demolition waste etc.

Bio-medical waste and industrial waste will be collected separately and will not be mixed with municipal waste. Further, bio-degradable waste shall be made use of. Such wastes have to be disposed of as per the procedure by the Bio-Medical (Management and Handling) Rules, 1998.

2. Segregation of Municipal Solid Wastes

In order to ensure *community participation* in segregation of waste, the municipal authority shall organise *awareness programmes* through *regular meetings* of representatives of local residents, welfare associations and non-governmental organization.

3. Storage of Municipal Solid Wastes

Municipal authorities are bound to establish and maintain storage facilities as they do not create unhygienic and in sanitary conditions. For this storage facilities shall be created keeping in view the quantity of waste generation and accessibility to users. It should not be open, aesthetically acceptable and user friendly.

Storage facilities or '*bins*' shall be 'easy to handle' design for handling, transfer and transportation. Bins for storage of bio-degradable waste shall of *green* colour, for recyclable waste with *white*, for other waste shall be painted with *black*.

Manual Handling of waste shall be prohibited, but if is unavoidable, it shall be carried out under proper precaution and with due care for the safety of workers.

4. Transportation of Municipal Solid Wastes

Vehicles for the use of waste shall be covered—not visible to public. It shall be prevented their scattering. Storage facilities—bins shall be cleaned on daily basis and should not be overflowed.

5. Processing of Municipal Solid Waste (Bio-degradable)

'Suitable Technology or combination' shall be used by municipal authority to *minimize burden* or *landfill.Bio-degradable wastes* shall be processed by composting, vermicomposting, anaerobic digestion or any other appropriate biological processing for stabilization of wastes. Separate procedure has also been provided by Schedule IV of the Rules.

Recoverable waste should follow the rule of recycling. Incineration, including pelletisation can also be used.

6. Disposal of Municipal Solid Wastes (Non-bio-degradable)

Only *non-biodegradable and other wastes which are not suitable either for recycling or for biological processing shall* be used for landfilling.

Schedule III of the Rules has also provided 'specifications for landfill sites'. It includes detailed rules for (i) site selection, (ii) facilities at the site, (iii) specifications for landfill, (iv) pollution prevention, (v) water quality monitoring, (vi) ambient air quality monitoring etc.

Objectives of the Study

- To study the existing MSWM practices in Agra
- To identify the bottlenecks in the present system
- To suggest a plan for MSWM in Agra in compliance with the MSW Rules 2000.

Methodology

The research approach for the case study is mainly qualitative

Primary and secondary Data collection: Data collection methods include document/literature review, semi-structured interviews, and observation. Information was gathered using a variety of methods to gain a better understanding of the situation, issues, perspectives and priorities. The primary data was collected by interaction with nearby residents, discussions with NNA (Agra Nagar Nigam) officials and other employees, taking photographs of the site. The major source of secondary data for this project was Nagar Nigam Agra, other government and non-government organizations and available literature

Present Scenario

This is an overview of the location, current extents of Agra city, its existing and projected populations, land use pattern and existing physical and social infrastructure. The information provided here was procured through various sources such as NNA, Agra Development Authority (ADA), discussions with the local residents.

Location

The city of Agra is situated on the Western Bank of river Yamuna at about 200 kms from Delhi in the state of Uttar Pradesh. The city is known for Taj Mahal, which draws tourist from all over the world. The city has an area of 141sq.km which falls under the jurisdiction of Agra Nagar Nigam. The city is divided into 90 election wards. These election wards are further grouped into 19 sanitary wards for solid waste management.

Population

As per as census 2001, the population of Agra is estimated 12.75 lakhs (excluding population of cantonment of Agra) with a decadal growth rate of 30.37%. The projected population for 2011 has been used in planning the collection and transportation equipment and the compost plant and the projected population for 2031 for designing the land fill site.

Physical and Social Infrastructure

- **Roads**

Agra is well connected by air, rail and road network. The city is considered important in view of tourism. The total road length maintained by Agra Nagar Nigam is 1820 km.

- **Strom Water Drains**

Nallahs or drains crisis is cross almost whole length and breadth of the city. Apart from the poor image of the city, these create unhygienic living conditions. The maintenance of the nallah's is poor and smaller nallahs have been encroached. This prevents cleaning and maintenance of the nallahs causing floods and overflowing in the heavy flow conditions mainly in the monsoons.

- **Sewage Drains**

The city of Agra has underground sewerage system, which is operated and maintained by the Agra Jal Sansthan. The city is divided into 25 sewerage zones, regrouped in 8 sewerage districts on the basis of topography and physical barriers like railway lines, river, National Highway etc.

About 50% of the sewerage system is not in working condition. Three Sewage Treatment Plants were built under Yamuna Action Plan Phase I. The STPs at Burhi ka Nagla (2.25 MLD) and Peela Khar (10 MLD) are made to perform beyond capacity, but still treat only 10% of the sewage they receive. Mostly the sewage goes into the open drains. The system is badly silted, choked and damaged.

- **Water Supply System**

Uttar Pradesh Jal Nigam (UPJN) is the nodal agency for developing water infrastructure in Agra. UPJN then transfers the facilities to the Agra Jal Sansthan (AJS) for operation and maintenance.

Agra water supply originates from river Yamuna. Apart from Yamuna Agra also receives raw water from Gokul Barrage. There are two water treatment plants at Jeevani

Mandi (Installed capacity of 200 MLD) and Sikandra (Installed capacity of 144 MLD).Both the treatment plants are giving a final output of about 270 MLD.

- **Electricity**

It is one of the most critical infrastructure facilities for a city. In case of Agra non-availability of power means people using DG sets as a back-up option. Agra city power supply has gone to private enterprise (Torrent powers).

The connected load has increased from 539 MW in 1991 to 627 MW in 2001. Currently with 57 substations of 33 KV, the available capacity is 493 MVA, which is just equal to the peak load demand of 500 MVA. However, keeping in view the growth of Agra, this needs to be increased.

- **Industries**

As per the report of District Industries Centre, Industrial Development in Agra at a glance, the total number of exporting Small Scale Industries (SSI) units is 131 and shoe export units are 87.

- **Health Services**

These include medical College which is providing medical facilities, 1 district hospital, and 197 private hospitals, about 101 government hospitals, 33 dispensaries, 21 maternity hospital, medical offices in the city, 15 child health care centers, 227 nursing homes and several private clinics. Besides these, there are hospitals for armed forces in the cantonment area, for police personnel in police lines and for railway staff in railway lines. Majority of the medical services are provided by private establishments.

Administrative Setup for MSW Management

The main organization which is responsible for urban governance and civic management is the Agra Nagar Nigam. The corporation has a democratically elected leadership from the constituencies within the geographic jurisdiction of the corporation boundaries. The present Corporation has been elected in the month of November, 2006 and has a jurisdiction of the corporation boundaries. Uttar Pradesh Municipal Corporation Adhiniyam, 1959 and its amendments govern the functioning of Agra Nagar Nigam (NNA). Under this cleanliness and sanitation is one of the main responsibilities of NNA.

The Municipal Commissioner (MNA) is the highest executive head of the Agra Nagar Nigam. The Health (Swasthya Vibhag) and Engineering (Nirman Vibhag) departments are the two executive departments responsible for implementation of the SWM plan and civil infrastructure in the town, respectively. The Health department is involved in planning and management of the SWM activities and in providing sanitary facilities to the residents of Agra.

The Engineering department works in coordination with the Health department and undertakes civil and infrastructure development activities for the city. The Nagar Swasth Adhikari and the Executive Engineer are the senior officers of the Health and Engineering departments respectively, who report to the Municipal Commissioner (MNA). The Nagar Nigam is divided into 90 election wards which are further divided into 19 sanitary wards being headed by sanitary inspectors. The sanitary inspector is responsible for managing the fleet routes, collection procedure, allocating diesel to vehicles, street sweeping and de-lineation of responsibility of the supervisors and safai karamcharis.

Existing MSWM system

The existing MSWM practices in Agra city. The various sources of waste generation, the current primary and secondary waste collection practices, waste transportation and disposal mechanisms, and gaps in the existing MSWM system are identified.

Waste Generation

The waste generation characteristics are influenced by climatic and seasonal variation due to tourist arrivals and festivals, the economy of the region, the physical characteristics of the city and social and religious customs of the city

The primary source of waste generation in Agra city is the local households, markets and commercial establishments such as hotels, restaurants, shops etc. Waste from industrial establishment, Hospitals and nursing homes, slaughter houses, street sweeping and sanitary drain etc. also contributes to MSW. Agra is famous for Petha sweets and petha making produces substantial quantity of organic waste. Construction and demolition waste, dairy waste also contributes to city waste generation. Apart from this there is a huge generation of leather and rubber trimming from small Footwear units spread in the city which adds up major portion to Municipal solid waste.

Major problem areas

- ***Waste being collected at Open Dumps and Dustbins***

Primary survey at secondary collection point i.e., Dhallao or open dumps sites and dustbins have been selected and detailed surveys have been carried out. As per NNA official reported there are 361 secondary collection points which includes 200 Open dumps/ Dhallao and 161 Dumper placer locations which were not actually found.

- ***Waste Reaching the Existing Dumping/Trenching Site***

The existing trenching site is located at Shadahra on Delhi-Tundla Bye pass Road. Waste transportation vehicles dispose waste at this authorized site from all over the city collected at the points designated by Nagar Nigam. The present efficiency of waste lifting to disposal site works out to be 75.5% of total waste collection at secondary collection points.

- ***Residential and Commercial Establishments***

The major portion of residents belonging to the residential establishments dump their household waste outside their homes and in some cases in dustbin (if available), or on the streets (open dumps) or in the drains running in front of the house. The karamcharis/operators collect waste from the various lanes and drains dump it in the vacant plots or in the open dumps. Private Operators is active in small way in MIG (Middle Income Group) and HIG (High Income Group) areas.

They lift waste daily from door to door at monthly charge of Rs. 20 to Rs. 50. This waste is transported to the secondary waste collection centers by Rickshaw trolley. A survey of the 50 residential establishments was carried out in order to assess the collection and transportation system at source.

- ***Industrial Establishments***

There are various small scale and cottage industries in Agra city. Based on observations made by the survey about these industries, the MSW is primarily generated from Petha and Footwear unit. Petha units are mostly located at Noori Darwaza and Raja Mandi .The petha waste is biodegradable and about 50% is eaten up by grazing animals. However significant amount of waste is thrown in nallah and nearby secondary collection points. The leather and rubber trimming from footwear industry generates significant waste. It was also observed that accumulated waste at open dump or dustbins was burnt in order to reduce the volume when it remains unattended for several days.

- ***Hospitals & Nursing Homes***

Apart from Bio-medical Waste generation, Hospital and nursing home generates noninfectious waste which contributes to total MSW generation.

Hospitals and Nursing home comprising both private and government are surveyed to analyze quantity and present Hospital waste management scenario in Agra city. As reported by NNA, the quantity of waste generated from these medical establishments is estimated at 1.49 kg/bed/day out of which Biomedical waste (BMW) and Non -infectious is 0.549 and 0.944 kg /bed/ day respectively. The value arrived is in line with the CPHEEO (Central Public Health and Engineering) guidelines.

- ***Hotels/Restaurants/Banquet Hall***

The waste from these establishments mostly includes left-over food, and disposable crockery. Workers of these establishments dump the wastes at nearby container/ open dump from where the same is transported to designated dumpsite by Nagar Nigam workers.

♦ ***Slaughter Houses***

There is one authorized slaughter house located at Chalesar, Kuber pur in Jamunapar Sanitary ward. The facts brought out during survey is that approximately 1200 animals (calves and buffaloes) are butchered every day in the peak season and 300 animals are butchered per day during the rainy season in the authorized slaughter house. Many other unauthorized location exist in Kajipara, Mantola, Sayedpara, Rawatpara, Khatipara etc. About 100 animals (calves and buffaloes) on an average in each of this unauthorized location. It was found that the slaughtering of small animals mainly lamb/goat/chicken is done at individual homes/ meat shops and the generated waste is disposed off in dustbins or open dump sites. The waste from slaughter houses (skin, horns, and bones) is sold to processing or recycling units, from where it is exported. The remaining waste such as flesh cuttings, gobar etc are thrown away to the points of Nagar Nigam at respective areas. No official records are being maintained on the quantity of waste generated from the slaughter houses. According to NNA waste from slaughter houses in Agra city is estimated to the tune of 7 MT/ Day per day. No official records are being maintained on the quantity of waste generated from the slaughter houses.

♦ ***Street Sweeping***

Street sweeping starts between 6:00 to 7.00 AM and continues up to 12:00 PM. The sweepers are provided with jharoo (brooms), pans, favda, hand-carts, panji (bamboo stick used to clean nalas [drains]), gayti (pointed favda to clean roads), and buckets to clean nalas. They collect the waste on road sides, which are then transported to the nearby open dumps/ dustbins by rickshaw trolley or to out skirt of the city directly by tractor trolley. Estimate of street sweeping waste were done at open dumps and dustbins, Based on the field survey, the street sweeping constitutes about 5% percent of the MSW generated in Agra.

- ***Drain Silt***

There are 25 major drains and 38 secondary drains with in the city. The major portions of the drain silt are generated by throwing of the household sweepings/ part of the street sweeping in the small drain flowing across the various streets and gullies. Some silt formation also happens through the blowing of the construction and other debris lying along side the drains. The drain silt typically comprises dust, household waste, sweepings, construction waste etc.

The drains are cleared of the accumulated solid waste twice or thrice a week/even fortnightly in a few parts of the city. At other parts, the waste is left unattended for months and cleared before the rainy season, up on complaints from the local residents.

The cleared waste is left for drying for one to two days adjacent to the drains (which again lead to a part of the waste getting back into the drain).

- ***Construction and Demolition Sites***

This waste varies from time to time depending up on the construction or demolition activities in Agra city. A major portion of this waste is generally used in reconstruction activities or for filling up of the low lying areas or kacha road. The individual generating construction waste generally engages private vehicle to collect the construction waste and dump it any elsewhere in the city for a nominal cost.

- ***Recyclables***

Rag pickers play a very significant role in keeping Agra City clean. The composition of recyclable waste comprises paper, textiles, plastics, Leather and rubber, glass and metal. Most of the value added recyclable is being sold directly to Kabadiwallas. The Rag picker community removes most of these recyclables.

MSW Collection System in Agra City

The MSW collection in Agra city is not well organized due to lack of awareness among the citizens as well as civic bodies responsible for collection of waste.

However, Collection conducted in a two stages. In first stage, the waste collected from door to door is transported to dustbins and open dumps. In this stage, collection is not very efficient even though large numbers of private sweepers are engaged in waste collection from door to door at a nominal charge.

In second stage waste filled DP Container is replaced with empty DP container by Dumper placer vehicle. The waste is transported to the designated dumpsite. Waste from open dumps is collected in trucks/tipper trucks/tractor manually or by JCB's and Loaders and finally transported to the designated dumpsite. The mode of transportation of waste from secondary dumpsites is decided on the basis of waste quantity as well as access road.

- ***Primary Collection Systems***

The primary collection of waste refers to house to house collection of waste in the community bins either by the resident themselves or by the sanitary workers. There is no organized arrangement for house to house collection of waste in almost whole city except for some part of the city. Community bins are also not available at convenient locations for depositing the waste.

- ***Secondary Waste Collection System***

The MSW collected from each of the primary collection points mentioned above is transported to designated open dump areas and DP containers (mostly on the main roads), which are the secondary collection points identified in Agra. Most of the waste is transported in rickshaw trolley and handcarts to the secondary collection points. The waste from the secondary collection points situated at congested places is lifted manually using pans and favdas tipper trolleys. In other cases JCB/ loaders are used to load the tipper truck/ trolley, which in turn are used to transport this waste.

The survey noted that the major secondary collection points are Open dumps, and DP Container in the city. All waste is finally dumped to Sahadra dumpsites. Waste is lifted in two to three trips per day by each vehicle. Also, vehicles used for transportation are not fully loaded with waste to the recommended vehicle capacity which means that the collection system has only 75-80% efficiency. We noted unauthorized dumping to near by low lying area by vehicle operators so as to save time and fuel.

MSW Transportation System in Agra City

This section describes the workshop for maintenance of transportation vehicles, the transportation equipment available with NNA, and the current MSW transportation practices followed in Agra city

- ***Workshop***

NNA has workshop located at Transport Nagar. The waste transportation vehicles are operated, supervised and maintained at this workshop. The workshop is supervised by workshop in charge. At present this responsibility is given to Junior Engineer. Based on the survey workshop includes one puncture department, one welding department, mechanical department and vehicle service station. Most maintenance and repair work is done here. Major Maintenance and repairs are outsourced on contract basis.

- ***Transportation Equipment***

Based on the survey, the NNA has total 96 waste transportation vehicles, used for the day-to-day collection of waste in the workshop. These vehicles make three to four trips per day for collecting and transporting the MSW from secondary collection points to the waste dump sites. These vehicles transport about 650 MT (Million Tonnes)/day of MSW.

The vehicles used for transportation of MSW start at the workshop in the morning hours and reach the sanitary ward offices located in each of the 19 sanitary wards. The sanitary inspector issues labor and fuel to each of these vehicles, when the waste collection process begins (8:00 AM). The vehicle drivers are supposed to make an entry in the log register, when they return in the evening to their respective sanitary ward offices. However, the entries are made in the log registers the following morning and that the sanitary ward offices are mostly closed by 12:00 PM.

- *Routing of Vehicles*

The existing unauthorized dump site is located at Shadhara on Delhi- Tundla bypass Road. The transportation vehicles make three to four trips per day to transport the MSW to this dump site. It takes about one to two hours to make three trips depending on the distance traveled and the traffic conditions.

- *Transfer Stations*

There are no transfer stations identified as part of the MSW transportation process. The waste collected from secondary collection points is directly transported to the existing dump sites within the city. However, the newly identified site at gram Raibha, Tehsil Kerawali on Bichpuri- Achenara Road is approximately 20 kms from the office of Agra Development Authority. If the waste has to be disposed at this site an intermediate station is required to transfer waste from small vehicles to large vehicle from where waste can be transported to this site.

Waste Processing and Disposal Mechanisms

This section describes the waste processing and disposal methods for each type of wastes adopted in Agra city. At present ANN does not possess any Waste processing Facility .The unsegregated waste is disposed at Shadhara dumpsite located at Agra –Tundla byepass road. Certain recycling waste is segregated and sold by Rag picking community. Various types of wastes were identified in Agra city - recyclables, bio-degradable wastes, and non-bio-degradable wastes

- *Recyclable Wastes*

The recyclable wastes are segregated manually by kabadiwalas and rag-pickers. The kabadiwalas purchase recyclable waste from residential and commercial establishments while rag-pickers collect recyclables from market places, dustbins, and dumping sites, and sort them before selling off.

- ***Bio-degradable Wastes***

The bio-degradable waste is not segregated either at the primary collection points, secondary collection points, or dumping sites. Most of the bio-degradable waste was found to be eaten by animals at the grazing on the open dumping sites.

- ***Non-biodegradable Wastes***

Street sweepings and drain silt is a major constituent of the non-biodegradable wastes. This type of waste is disposed off at the dumping sites along with other wastes without any prior processing.

Status of Present Disposal & Proposed Landfill Site

The existing SWM system for Agra does not have an engineering landfill site for disposal of waste. The waste collected from secondary collection points is dumped in an unorganized manner at the dumpsite. The authorized dumpsite is located at Shadhara on Agra- Tundla Bypass Road. Behind this site Kalindi Vihar area is situated and on the other side Jharna Nallah.It is at a distance of 4-5 kms from Workshop at Transport Nagar.

The present situation of the landfill site is over dumped to a height of approx 8 ft from Road level. The total area of the Dumping site is 4.475 acres. The way to the dumping site is badly damaged, due to which the vehicle of Nagar Nigam for waste disposal transport waste less than their capacity. During the discussion with NNA, it was also brought out that situation become worst in rainy season as the way becomes to much slippery due to which there exists risk of frequent accident.

The Agra Nagar Nigam has 22 acres of land at kuberpur, teshil, Etmadpur, which has been proposed for the development of Integrated Sanitary Landfill Facility.

Gaps in Existing MSW Management with Respect to MSW Rules 2000

The *MSW Rules 2000* has laid out very detailed and specific criteria for management of solid waste. It has also laid down the guidelines, guidance notes and specific work standards to be adopted by municipal personnel or personnel involved with solid waste management for organized SWM practices (starting from waste collection, segregation, transportation, to environmentally safe waste disposal practices).

Some of the significant gaps observed by us are as under:

- ***No Segregation Practiced at Source***

The waste generators in Agra city do not segregate the waste prior to disposal. It is identified that these generators (households, restaurants, shops, etc.) dump mixed waste into dustbin, drains, and open dumps

- ***Primary collection of solid waste is not appropriate***

NNA does not provide door-to-door MSW collection service to all its residents the major portion of HIG and LIG residents of Agra city pay private operators to collect their daily waste in handcarts and disposed off to near by secondary collection point. However, others throw the household waste outside their residences from where sweeper trolley and dump the same into the dustbins or onto streets (open dump). In most of the cases, generators themselves dispose off waste in nearby waste collection points/ containers, onto the streets, or in the nearby drains.

The *safai karamcharis* employed by the NNA do street sweeping, collect drain silt and waste heaps from roadsides and dispose them off at a nearby open dumps. These unorganized disposal methods have resulted in accumulation of solid waste on roadsides and vacant plots and in low lying areas and storm water drains.

Door-to-door waste collection service has to be provided to households. The roadside waste collected by street sweepers must be directly dumped into a separate bin at the secondary waste collection point.

- ***Secondary Storage of Solid Waste is Unorganized***

We observed that at places wherever dustbins are available, either they are rusted or damaged. At other places, waste is dumped on open dumps which have evolved over time. In the absence of secondary storage facility for MSW, it is dumped at any location in the vicinity - drains, vacant plots, street corners, low lying areas, and other open areas. Heaps and stretches of un-segregated waste in open areas is an eyesore, thereby causing environmentally hazardous & unhygienic conditions across the city, thus, creating conditions for breeding of mosquitoes, grazing by cattle

Separate coloured bins must be provided at the secondary storage location for biodegradable and non-biodegradable, and recyclable wastes. The bins must be covered and cleared at the scheduled time to prevent storage of waste for a long time and littering of waste outside the bins.

- ***Solid Waste is Transported in Open Vehicles***

Most of the times, solid waste is transported in open trucks and trolleys. We observed that these vehicles are overloaded with waste, resulting in road littering during transportation. The loading and unloading of waste is done manually and safai karamcharis involved in this activity do not use any Personal Protection Equipment (PPE) for their protection.

The waste transportation vehicles must be covered at all times expect while loading and unloading activities and the loaded waste should not exceed the capacity of these vehicles.

- ***Slaughter House Waste is Mixed with the MSW***

Waste from slaughter houses is dumped along with the MSW in open and near by agricultural land. As such, there is no provision for segregation and safe disposal in Agra city. Sweepers of NNA collect waste by means of rickshaw.

Slaughter house waste should be collected separately and disposed off by controlled incineration, burial, anaerobic digestion, or other approved processing methods.

- ***Bio-medical waste is not managed properly in all health care facilities***

The study found out that most of the private hospitals/ nursing homes in Agra are segregating their waste and the bio-medical waste. They have contracted with a UPPCB Uttarpradesh Pollution Control oard) approved contractor M/s Dutta Enterprises of Agra for collection and disposal of BMW. However few organizations do not strictly follow segregation of BMW Bio-medical Waste) and needs to be penalized by the competent authority.

Bio-medical waste segregation, handling and disposal mechanisms adopted by all medical institutions must be strictly in compliance with the Bio-medical Waste Rules.

- ***Collection and disposal of construction waste is not appropriate***

The construction/ demolition waste generated by local residents is transported in tractor trolleys and disposed off in open/ low-lying areas in the vicinity, privately.

The construction and demolition waste (from private or NNA sites) must not be dumped in any open areas in an unorganized manner. It must be handled under the guidance of NNA staff.

- ***Disposal of Solid Waste is not Appropriate***

The solid waste collected from various sources is disposed off in open dumpsites indiscriminately without segregation or preprocessing. There is no engineered sanitary landfill site for safe disposal of solid waste.

According to MSW Rules 2000, biodegradable waste shall be processed and converted into compost or used for power generation; recyclables shall be segregated and sold to recyclers; no hazardous waste be dumped along with MSW; construction waste to be segregated and used for filling low lying areas and only remaining waste shall be dumped into engineered landfill facility.

- ***Manual Handling of Solid Waste***

Safai Karamcharis involved in primary collection of MSW do not use any Personal Protection Equipment (PPEs) such as face masks, disposable gloves, boots, hats, and proper safety clothing (sturdy colored uniform) to avoid direct contact with waste and reduce the likelihood of on-the-job injury. Manual handling of solid waste during primary collection is an acceptable practice in Agra city.

- ***Lack of Awareness Among City Residents and Civic Authorities***

The NNA staff is responsible for managing MSW in Agra city in accordance with the *MSW Rules 2000*. However, they are completely ignorant of these rules and the implementation practices recommended in the document. Public participation is very essential in successful implementation of the MSW management plan in the city. Therefore, a planned and concerted effort is required to bring about awareness among the public and make them realize their responsibilities as individuals and as a community. In summary, public awareness, community participation, transparent administration, accountability at all levels is the need of hour so as to ensure success of any MSW management plan.

Proposed MSW Management System

The proposed SWM system is broadly based on the *4R Environmental Protection Rules* (Reduce, Recycle, Reuse, and Recover) and is in accordance with the *MSW 2000 Rules*. The primary aspects of the proposed plan include the following:

- Compliance with MSW Management & Handling Rules of 2000
- Waste to be covered at all stages of handling
- Elimination of manual handling of waste, and the provision of the proper PPEs to the workers
- 100% collection and transportation of the generated waste

- Promote information, education and communication across the stakeholders to ensure system efficiency and sustainability
- Adequate health and safety provisions for workers at all stages of waste handling
- Regular environmental monitoring at waste processing and disposal facilities
- Have robust complaint-handling system in place
- Conduct regular internal and external independent audits on the efficiency of entire SWM system

Proposed MSW Management Plan

This section describes a step-wise approach for safe and efficient ***collection, storage, and transportation*** of MSW for Agra city. They include primary waste collection, secondary waste collection and storage, and waste transportation for subsequent processing and disposal.

Primary Waste Collection System

The efficiency of the proposed waste management plan described below is driven by the separation of waste at the primary collection level. For this purpose, following approach needs to be adopted by the residents as well as the NNA personnel:

- The waste should be stored by the generators in two separate bins, one each for biodegradable and recyclables.
- Door-to-door waste collection should be privatized in the entire Agra city. Area wise responsibility will be assigned to private operators. Waste collectors should collect waste on a day-to-day basis in two types of bins - green bins for bio-degradable and white bins for recyclable waste. Door-to-door collection of waste should be through rickshaw trolleys. Initially NNA safaikaramcharis should do street sweeping on a day-

to-day basis. The street sweeping waste should be collected in wheelbarrows. The waste collectors should transfer bio-degradable and non-biodegradable waste to the secondary collection location and should sell the recyclable waste.

- The existing infrastructure for secondary collection of waste should be upgraded to comply with MSW Rules 2000 and additional secondary collection locations we should be developed to ensure that no primary collection staff has to travel more than 250m for dumping of waste.
- The NNA safaikaramcharis should re-use the construction and demolition wastes for filling-up the man-made low-lying areas or shall recycle the construction waste to make bricks. The local residents should hand-over the construction waste generated from private construction activities to NNA workers for proper use.
- Most of the waste generated from slaughter houses such as bones, skin and horns is recyclable and should be sold to the recyclers. The only solid waste generated from the slaughter houses is flesh cuttings, which should be buried in a designated area.
- The private workers and NNA safaikaramcharis should be well equipped to avoid direct contact with waste.
- It is proposed that primary waste collection should be done by NGOs or private operators. Presently, in Agra there is limited system of primary collection of waste. It is proposed that NNA shall identify NGOs/ private operators for this purpose and authorize them for undertaking this activity.

- ***Collection of Waste from Households***

The door-to-door collection of waste should be done on a day-to-day basis between 7:00 AM and 2:00 PM. The NNA should ensure that infrastructure is made available for undertaking this activity in compliance with the MSW Rules 2000.

- ***Required Infrastructure for Households***

The proposed infrastructure for primary waste collection is given below.

The infrastructure is calculated at ward level. Based on field survey, it is proposed to use Rickshaws for door-to-door collection of waste. It is proposed to have a Rickshaw with four 50 litre bins for door-to-door collection of waste. Among 4 bins, 3 bins will be used for collection of bio-degradable waste, 1 bin for recyclables.

The distribution of bins for waste collection has been done on the basis of the waste characteristics received from the households. The waste received from the households mainly comprises of bio-degradable waste. The household waste also contains recyclable waste but the quantity of this waste as compared to bio-degradable waste is very less.

Presently handcarts are used for primary collection of waste. Out of the total existing handcarts, 75 handcarts are in good condition but these handcarts do not facilitate collection of segregated waste.

- ***Required Tools and Equipments***

Each worker will be provided with a broom and a dustpan for waste collection. The workers involved in SWM also need to be equipped with adequate personal protective equipments (PPEs). Each worker engaged in door-to-door collection of SW shall be provided with gloves, boots and uniform. A uniform for a worker will include two sets of uniform for summers and two sweaters for winters.

- ***Collection of Waste from Commercial Areas***

The association of markets, shops and establishments should organize for primary collection of waste with the help of private waste collectors authorized in their area.

Most of the waste from commercial areas is recyclable. Therefore it is proposed that rag pickers should collect recyclable waste from shops and establishments as soon as they open. There should be a secondary waste collection point in every market area; therefore shopkeepers themselves shall dispose off their waste in these collection points.

The usual timings of waste collection are not being feasible for markets as shops normally open after 9am. It is proposed that street sweeping in market area should be done in the early morning hours and waste collection shall be done from 9:00 am to 10:00 am.

- ***Collection of Construction and Demolition Waste***

The construction and demolition waste should be collected on need basis. It is the responsibility of generator to inform NNA that construction waste need to be collected from the site. The NNA should collect this construction waste in existing tractors and trucks and transport it for reuse or recycle. The NNA should identify low lying areas for dumping of this waste. Existing trucks can be upgraded and can be used for this activity.

- ***Collection of dairy waste***

All dairy waste should be sent to the compost plant. Existing tractors with NNA can be used for this purpose. The compost plant operator should have understanding with dairy owners and NNA for collection and transportation of dairy waste from source to the compost plant site and transportation cost would be reimbursed by the Compost Plant operator

- ***Collection of waste from banquet halls, marriage halls and community centers***

It is proposed to deploy a few auto rickshaws for collection of bulk waste from these facilities. The facility owners' have to request NNA to get this waste collected from their premises. NNA shall charge them for providing this facility. Apart from this, NNA will have tie-up with private vehicle operators and NNA will use these vehicles to lift the waste from any such site.

- ***Street Sweeping***

The street sweeping should be done on a day-to-day basis. It is desirable to split the 8 hours of duty of sweepers into two shifts, 4 hrs in the morning and 4 hrs in the afternoon. The street sweeping is to be done by NNA safaikaramcharis as per the proposed plan.

It is proposed that approximately 25% of the roads which are prime roads remain in good condition all round the year should be cleaned by road vacuum sweeping machine. Initially it should be done by safai karamcharies. Over time it will be shifted to private operator in phased manner.

Secondary Waste Storage System

There are open dumps, dhalaos and dumper containers for secondary waste storage in Agra. The existing collection system is taken as a basis to develop secondary waste collection system for Agra city.

- ***Proposed Management Plan for Secondary Waste Collection System***

There are two types of waste streams from the primary collection of waste – 'bio-degradable', and 'recyclables'. The recyclables should be sold directly in the market by the waste collectors.

It is proposed to provide secondary collection facility for bio-degradable and non biodegradable waste. MSW at every secondary collection point shall be stored in two separate covered containers – green for bio-degradable and black for non bio-degradable.

Non-biodegradable will include drain silt and street sweepings. The sweeper shall not travel more than 250m to dispose off primary collection waste. Therefore distance between two secondary waste collection locations shall not be more than 500m.

In wards with leather industry, provision shall be there for separate collection of leather waste.

The bin design and strength shall be able to facilitate its hydraulic lifting by transportation vehicles. The secondary waste collectors shall be well equipped to avoid direct contact with waste.

- ***Proposed Infrastructure for Secondary Waste Collection System***

The area under NNA is approximately 190 sq. km. It is proposed to have one secondary collection point for every 400mx400m area in high density wards and one secondary collection location for every 500mx500m area in remaining wards of NNA.

It is suggested to have Waste infrastructure Depot (WIDs) in each sanitary ward at prime locations.. The external façade of these WIDs will also be used for advertisements to generate revenue.

Two types of infrastructure are proposed for secondary collection of waste - dumper placer containers and refuse collector bins. At locations where expected biodegradable waste is more than 2 m3, it is proposed to have 3.5 m3 capacity dumper placer container, locations where expected waste is more than 3.0 m3 it is proposed to have 4.5 m3 capacity DP container and locations where waste is more than 4.5 m3 it is proposed to have 8.0 m3 capacity DP container. The points where expected waste is less than 2.0m3, it is proposed to have adequate number of 1.1m3 and 0.6m3

Refuse Collector bins depending upon the expected quantity.

It is proposed that DP containers shall be kept at 94 locations and on remaining 464 locations Refuse Collector (RC) bins shall be placed. Two types of RC bins are proposed - 1.1m3 capacity and 0.6 m3 capacities. 1.1 m3 capacity bins are proposed at 200 locations and 0.6 m3 capacity RC bins are proposed at 264 locations.

The above secondary collection plan is developed keeping in view the available infrastructure with NNA. It is proposed that existing Dumper Containers available with NNA shall be used in the new proposed plan. There are two types of DP containers available with NNA - 8 m3 and 4.5 m3 capacity. It is proposed that with time when existing DP containers worn out, they shall be replaced with 3.5 m3 DP containers as per the design.

Placement of DP Containers

It is proposed that DP containers will be placed at 94 locations. At each location pair of DP container will be provided- one for bio-degradable waste and one for non biodegradable waste. It is proposed to use existing dumper containers at 75 locations. It is proposed to buy new 3.5 m3 twin container dumper containers for the remaining 19 locations.

Placement of RC Bins

Apart from the above, additional 150 litre twin bin set shall be placed near all public places including schools, colleges, offices, post office, and market areas. It is proposed to provide 270 pairs of 150 litre bins for placement across the city.

References

Ministry of Environment and Forests (MoEF) (2000) " Municipal Solid Wastes (Management and Handling) Rules,Ministry of Environment and Forests, Government of India, New Delhi.

Regional Centre for Urban & Environmental Studies, "Project Report for Solid Waste Management in Agra, Uttar Pradesh".

Tchobanoglous G, Hilary Theisen, Samuel A. Vigil (1993) "Integrated Solid Waste Management - Engineering Principles and Management Issues", McGraw Hill, International Edition.

Central Pollution Control Board (CPCB) (1999) "Status of Solid Waste Generation, Collection, Treatment and Disposal in Metro Cities "Central Pollution Control Board, Delhi.

CPHEEO (2000),"Manual on Municipal Solid Waste Management", Central Public Health and Environmental Engineering Organisation, Ministry of Urban Development, New Delhi.

http://www.inext.co.in/epaper

http://www.indianexpress.com/supplement/India/798/

http://www.nlsenlaw.org/live/nlsenlaw/agriculture

http://localbodies.up.nic.in/swm/chap2.pdf

http://www.cpcb.nic.in/Municipal_Solid_Waste.php

http://www.nswai.com/databank.php

http://www.nagarnigamagra.com/disposal.htm